The Elements of Information Gathering

The Elements of Information Gathering

A Guide for Technical Communicators, Scientists, and Engineers

by
Donald E. Zimmerman
and
Michel Lynn Muraski

Oryx Press
1995

The rare Arabian Oryx is believed to have inspired the myth of the unicorn. This desert antelope became virtually extinct in the early 1960s. At that time several groups of international conservationists arranged to have 9 animals sent to the Phoenix Zoo to be the nucleus of a captive breeding herd. Today the Oryx population is over 800, and nearly 400 have been returned to reserves in the Middle East.

4041 North Central at Indian School Road
Phoenix, Arizona 85012-3397

Published simultaneously in Canada
Printed and Bound in the United States of America

♾ The paper used in this publication meets the minimum requirements of American National Standard for Information Science—Permanence of Paper for Printed Library Materials, ANSI Z39.48, 1984.

Library of Congress Cataloging-in-Publication Data

Zimmerman, Donald E.
The elements of information gathering : a guide for technical communicators, scientists, and engineers / by Donald E. Zimmerman, Michel Muraski.
Includes bibliographical references and index.
ISBN 0-89774-800-X
1. Science—Information services—Methodology—Handbooks, manuals, etc. 2. Engineering—Information services—Methodology—Handbooks, manuals, etc. 3. Information storage and retrieval systems—Science—Methodology—Handbooks, manuals, etc. 4. Information storage and retrieval systems—Engineering—Methodology—Handbooks, manuals, etc. 5. Communication in science—Methodology—Handbooks, manuals, etc. 6. Communication of technical information—Methodology—Handbooks, manuals, etc. I. Muraski, Michel Lynn. II. Title.
Q224.Z56 1995
001.4'33'0245—dc20

94-36806
CIP

CONTENTS

LIST OF ILLUSTRATIONS

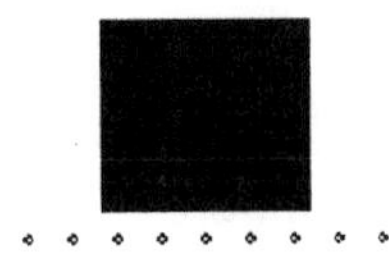

Figures

Panels

Tables

PREFACE

As a practicing technical communicator, scientist, or engineer, or as an academic major in one of these fields, your success depends heavily on your information-gathering skills. Good information-gathering skills can help you solve problems, focus your inquiry, generate alternative approaches to problems, save you time and money, keep you abreast of your field, and help you investigate other fields.

Whatever your profession or academic major, refining your problem-solving skills will play a major role in your success. Key to developing your problem-solving skills are efficient and effective information-gathering skills. Good information-gathering skills enable you to focus and refine problems, generate alternative approaches to problems, and identify information needed to solve problems.

Good information-gathering skills also save you time and money. By knowing where and how to search for information, you can expediently identify key sources, locate specific information and retrieve it with the least effort and expense.

Technical communication and scientific and engineering fields change rapidly, and good information-gathering skills will enable you to keep abreast of the constant changes. By being able to identify and retrieve the needed information, you can learn of recent developments and advances in your field. Furthermore, professionals today are frequently called to investigate topics in other fields, and good information-gathering skills will enable you to learn about related fields.

Developing good information-gathering skills is not easy. Why? Because information expands exponentially, you must winnow only the appropriate information sources—whether printed, electronic, or people—from the proliferation of sources.

The Elements of Information Gathering will help you develop your information-gathering skills. Part 1 provides an overview of the scientific method and problem solving, and then explores library searching and suggests strategies for managing information; part 2 discusses individual and group interviewing and details the survey process; and part 3 suggests strategies for advanced research methods.

In part 1, chapter 1 recommends that you consider using a scientific methodology for gathering information, and chapter 2 suggests that you appraise the logic of problem solving to enhance the information-gathering process. Chapter 3 presents strategies for learning library resources. Chapter 4 identifies key printed resources and then provides strategies for searching those resources. Chapter 5 identifies the emerging electronic and online resources and offers you an approach for conducting electronic searches and navigating information highways. Chapter 6 gives you practical advice on retrieving and evaluating printed and electronic literature and suggests ways to manage that literature more effectively.

In part 2, chapter 7 provides you with the foundations of conducting informational, one-on-one interviews, and chapter 8 explains how to conduct focus groups and the nominal group technique. Chapter 9 provides you with an overview of the survey process, and chapter 10 explains how to identify and sample groups. Chapter 11 tells you how to develop questionnaires or survey instruments. Chapter 12 provides the foundations of administering a survey, and chapter 13 explains the basics of analyzing and interpreting your survey data.

In part 3, chapter 14 provides a basic overview of conducting usability evaluations. Chapter 15 details how to conduct case studies and explains the fundamentals of ethnographic research. Chapter 16 provides a brief overview of advanced research methodologies required for conducting experiments.

Appendix A discusses accessing information through the Internet, and Appendix B reviews bibliographic software and its uses in managing information and writing papers.

ACKNOWLEDGMENTS

We sincerely appreciate the assistance of Herbert Michaelson, acquisitions editor and long-time Society of Technical Communication member, who encouraged us to submit our book proposal to Oryx Press.

At Oryx Press, we acknowledge the assistance and guidance of Art Stickney, Anne Thompson, and Sean Tape.

We recognize the following professionals who played key roles in helping us develop the information-gathering skills that formed the foundation of this book: Robert J. Robel, Thomas Haig, V. E. Suomi, and Bud Sharp.

We thank the librarians at Colorado State University's Morgan Library: Joan Chambers, Antoinette Lueck, Barbara Burke, and Fred Schmidt. We give special merit to the professional seminars on electronic data searching presented by Barbara Burke, Anna DeMiller, Larry Rouch, Evelyn Haynes, Elizabeth Fuseler, Minna Sellers, and Michael Culbertson. We thank Abigail A. Loomis, Coordinator, Library User Education Program, University of Wisconsin-Madison, who provided background materials and information on the University of Wisconsin's library and its user education program.

Don sincerely appreciates his family's tolerance of his continued writing pursuits. He is indebted to his wife, Marietta, and children, Rachel and Jeramy, for their continued support.

Michel enthusiastically acknowledges Jim, for his love and equanimity, and her beloved dog, Custer, for his long-suffering vigilance.

PART 1

GATHERING AND EXPLORING INFORMATION SOURCES

CHAPTER 1

Extracting Information with Scientific Methodology

The passage through profuse information to prodigal resources to useful references may seem like a walk through a mist. You most likely have experienced the anxiety of trying to extract specific, salient references for your research from the proliferation of information available and realize that the problem is not so much the availability of information, but the overload. What you need is some way of ferreting out and organizing subject-specific data toward a purposeful end.

Consider Samantha, a graduate student who returned to school after several years as a professional. In September of her first semester, Sam lamented to her professor that she had found only two articles by a well-known researcher in the technical communication field. She could not understand why he was so celebrated when he had apparently published only two articles. Her professor was considerably puzzled since he had read several of this researcher's articles and knew that the researcher had not only published many articles but also a book. After questioning Sam about her search strategies, the professor identified Sam's problem—she'd been searching for leads in all the wrong places.

The professor knew that Sam's well-known researcher did not publish in the journalism and mass communication literature but rather in the human-computer interaction, computer science, and technical communication literature. The professor verified this when he logged into the university's network and searched the library's online catalog and periodicals database. Within five minutes, Sam's professor identified a dozen titles by the author. Apparently, Sam had not tried the online catalog and periodicals database.

One month later, Sam told her professor that she had driven 60 miles to another research university library because the requisite information was not at her university's library. This was a different, but ancillary, problem—by searching in all the wrong places, Sam conducted neither an accurate nor an efficient examination. The professor advised Sam that she could save time, gas, and money by using the university's interlibrary loan service or electronic document delivery service.

Sam's predicament is familiar to many undergraduate students, graduate students, and professionals who, lacking adequate library and information-seeking skills, waste time and money and become extraordinarily frustrated. Sam was thwarted primarily because she did not know how to conduct an efficient search through the more than 1.25 million volumes in her university's research library.

She's not alone.

PROLIFERATION OF INFORMATION

Technical and scientific information abounds and multiplies exponentially.

The R. R. Bowker Company, which has provided statistics on the book industry since 1880, reports that in the United States alone total book production has expanded from 2,076 titles in 1880 to approximately 46,743 in 1990 (Peters 1992). Likewise, academic research journals now include journal collections such as EMBASE (Excerpta Medica journal collection), which reviews about 3,500 biomedical scientific journals (Elsevier Science Publishers 1989), and Index Medicus, which indexed 3,058 biomedical health science journals as of January 1993 (National Library of Medicine 1993). The indexing branch of the National Agricultural Library and its cooperators currently index 2,012 journals for AGRICOLA, the AGRICultural OnLine Access, the library's bibliographic database (Dowling and Lehnert 1992). In 1989/1990, PIE (Publications Indexed for Engineering), an Engineering Information, Inc. collection, abstracted and indexed 1,900 publications (including journals and conferences). In addition, Engineering Information, Inc. identifies an additional 200 publications that are indexed selectively because the publications are multidisciplinary and cross-disciplinary journals or magazines.

To successfully extract technical and scientific information from all available sources, you will need an understanding of the scientific methodology. In contrast to other ways of learning about the world, science has fundamental characteristics that engender the focus and precision necessary in the information-gathering endeavor.

SCIENTIFIC METHODOLOGY

According to Babbie (1992), you will most likely stumble if you set out to learn about your world through casual inquiry. Babbie offers that scientific inquiry, based on logic and valid and reliable observation, provides safeguards against common, casual inquiry.

Babbie suggests that these casual, information-gathering errors include inaccurate observation—failing to observe the obvious while mistakenly observing the nonexistent; overgeneralization—assuming that a few, similar events are evidence of a larger, general pattern; selective observation—paying attention to those events that correspond with your overgeneralization and ignoring those that do not; made-up information—inventing information to explain events that contradict your overgeneralizations; illogical reasoning—brushing away contradictions with strokes of illogic; ego involvement—which leads you to create a formidable barrier to further inquiry and accurate understanding to avoid disproof of your understanding; and premature closure of inquiry—when all the aforementioned conspire to end inquiry too soon.

A scientific understanding of the world, in contrast, must make sense and correspond with what you observe. These elements are essential to science and relate to three major aspects of the scientific enterprise: 1) theory, the logical aspect of science; 2) research methods, the observational aspect of science; and 3) statistics, the device for comparing what is logically expected with what is actually observed.

While specifics of a scientific methodology will vary with the scientific discipline, most scientific methodologies include the following characteristics:

- Understanding the problem
- Stating the problem
- Collecting data
- Analyzing data
- Interpreting data
- Drawing conclusions

Understanding the Problem

Researchers must develop an understanding of the problems in their field. Many researchers spend years, even decades, backgrounding themselves on the problem they are investigating. They learn to think in ways that are unique to their discipline, often called a scientific paradigm, which includes the language, conceptual framework, theories, methods, and limits of their discipline. They read and review relevant and associated literature to provide themselves with insights into their problem. They talk with other scientists and spend hours thinking about how to investigate their problem.

The scientific methodology of understanding the problem is, however, more than a fact-gathering activity. Without some guiding idea, researchers will not know what information is relevant and what is not. Kerlinger (1986) believes that while it is not often easy for a researcher to formulate the problem simply, clearly, and completely, an adequate understanding of the research problem is one of the most important components of research. According to Kerlinger, if you want to solve the problem, you must generally know what the problem is.

Stating the Problem

A good problem statement drives scientific investigations. Depending upon the scientific field, the problem statement may be called an hypothesis, a research question, a problem statement, or an objective. Researchers conducting basic research often use theoretical perspectives to drive research hypotheses, which they then test in field and laboratory experiments. These methods produce general explanations, or overviews. Researchers conducting applied research to solve a specific problem use literature reviews to derive purposeful statements and objectives that direct their investigations.

All problem statements direct basic and applied researchers to look at narrowly defined problems. Such problem statements influence the research methods selected, the kind of data collected, the nature of the data analyses, and, potentially, the conclusions. Posing problems properly, however, is often more difficult than answering them.

Kerlinger (1986) identifies three criteria of good problem statements. First, the problem statement should express a relation between two or more concepts, such as, how is A related to B under conditions C and D? Second, the problem statement should be clear, unambiguous, and in question form because questions have the virtue of posing problems directly. Third, the problem statement should be such as to imply possibilities of testing. Kerlinger contends that a problem statement that does not contain implications for testing its stated conceptual relations is not scientific.

For example, suppose you work for a company that is considering whether an employee wellness program would save it money. You might pose the following problem statement: "Will implementing a wellness program be cost effective for XYZ company?" Although this question captures the essence of the issue, it does not address the issue with the specificity needed to conduct a focused investigation of the problem. What do you mean by a "wellness program"? What do you mean by "implementing"? What do you mean by "cost effective"? To pose the kind of problem statement that guides a focused, productive research, you must now identify, with greater clarity, the problem statement's concepts.

Examine the concept of "wellness program." "Wellness program" might include 1) conducting workshops on diet, 2) conducting workshops on appropriate exercise, 3) providing exercise facilities and locker rooms, or 4) conducting workshops on smoking and drinking cessation and stress management.

Next, consider that the concept "implement" could carry several meanings, including 1) the company provides a conference room for the workshops, but employees must pay the costs of trainers, 2) the company provides workshop space and hires trainers, or 3) the company provides the workshop space, hires trainers, and pays incentives for employee participation.

Finally, explore the concept of "cost effective": 1) How much will these programs cost? 2) What will the payback period be? 3) In what ways might the payback be measured? 4) Will savings include reduced sick days or reduced workers' compensation claims?

As you reflect on the initial problem statement and its concepts, you will be able to generate more subject-specific questions. Then, after a thorough examination of each concept, you are better prepared to consider a more narrowly stated problem. You now ask: "Will investing $75,000 a year in a wellness program reduce costs due to employee sickness and injuries?" This problem statement begins to provide guidance for directed exploration.

Collecting Data

After specifying the concepts to be studied in your problem statement, you must choose a research method and create concrete measurements. A variety of research methods exist, although each field often has unique data collection techniques. Technical, scientific, and specialized professionals gather information through 1) reviewing literature, 2) talking with other professionals, and 3) observing factors associated with the problem.

While reviewing literature is often a powerful method of extracting what others have done, finding the relevant information to review becomes difficult in today's information-rich society without a systematic method. Talking with others is also more productive if you use the systematic approaches of structured interview and survey techniques. Observation also has a wide range of systematic, structured approaches, which may include direct observation, experiments, tests of people's reactions, tests of equipment, or participant observation.

For example, climatological researchers often directly observe characteristics of weather phenomena. This type of research is classified as field research. Climatologists may also conduct laboratory experiments, commonly known as experimental research. These same researchers may, on occasion, interview experts about weather phenomena. This type of research is classified as survey research. Finally, climatologists may systematically conduct a content analysis of historical documents to look for patterns, to understand weather fluctuations over the years, and to understand possible weather changes. This method is called interpretative research.

After selecting the most appropriate research method, you must move from vague conceptualizations, or ideas, to concrete measurements of the problem under analysis. Most researchers use the process of conceptualization to refine and specify abstract concepts into precise terms. For example, assume your problem statement is, "Does education reduce prejudice?" First, you should identify the concepts in that statement as "education," "reduce," and "prejudice." Next, you should identify measurable observations of "education," "reduce," and "prejudice." Thus, your observable measurements for "education" might be "cultural awareness seminars," "diversity literature," or "ethnicity presentations."

When researchers construct and evaluate measurements, they pay particular attention to two fundamental characteristics of measurement quality—reliability and validity. Reliability is the consistency and repeatability of measurements over time. You must ask yourself if your measurement procedure yields the same description of a given phenomenon if that measurement is repeated. Validity refers to the extent to which a measure reflects the concept under investigation. You must query the extent to which a specific measurement provides data that relate to commonly accepted meanings of a particular concept. Many internal and external threats to validity exist for unwary researchers. (See chapter 16 for further explanation of these concepts.)

Analyzing Data

Researchers use a variety of methods to analyze their data. Biological and physical scientists frequently use mathematical and statistical analyses to help them make decisions about their data. In some scientific fields, such as geology and taxonomy, scientists provide narrative descriptive illustrations of their observations. Social scientists may use qualitative and quantitative methods for analyzation. Qualitative methods usually entail narrative descriptions of events or observations while quantitative methods usually involve statistical and other data analysis techniques.

Interpreting Data

With data analyses completed, researchers interpret the data in light of their problem statement. When the results are as hypothesized, scientists cautiously put meaning on their data. When hypotheses direct the research, the careful researcher either rejects or fails to reject hypotheses. The study can lend support to the hypotheses, but never proves them completely. As more studies lend support to the specific scientific subject, more support amasses for the hypotheses.

When data do not support the hypotheses, this too provides understanding of the particular problem. Knowing that something does not work, or that no relationship exists, can often help scientists rethink a particular perspective, gain new insights into a topic, or identify alternative explanations for occurrences. When interpreting data, researchers also explore the aberrations or the unusual findings of their studies, which may also provide insights that will help further the researchers' understanding of the particular problem.

Drawing Conclusions

Based on the problem, data collection, analyses of the data, and data interpretation, researchers draw conclusions about the particular problem. In drawing conclusions, they try to present their findings succinctly and identify areas of further study. In many cases, researchers can summarize their findings into one or two key points. At this point, researchers often identify spin-off topics for further research and investigation. Whether these ideas come from the primary results of the project or the aberrations, spin-offs frequently provide the foundation for subsequent research.

COMMUNICATION SCIENCE

In 1963, Wilbur Schramm contended that communication was not a scientific discipline like psychology or mathematics but rather an academic crossroads where many passed but few tarried (Schramm 1963, Berger and Chaffee 1987). A generation later, Berger and Chaffee (1987) offer that so many academic traditions tarried at this crossroad that a large urban center developed. This figurative city—communication science—is grounded in the traditional scientific method of inquiry; its cornerstones are theory, operationalization, observation, and generalization; it is constructed of diverse research methodologies; and it is inhabited by interdisciplinary families, including most humanities.Through scientific inquiry, communication science seeks to understand communication phenomena such as messages, signals, and symbols, and their effects in various contexts, including nonverbal, interpersonal, organizational, and mass.

The concept of a science of human communication assumes that human behavior can be both understood and improved through systematic study. This assumption, of course, applies to all science, but it may be a barrier for you when you first consider communication science. Certainly, it would appear initially that the physical sciences are more predictable than communication science. For example, water will melt at and freeze at 32° F with great predictability. A person, however, may send the same nonverbal signal at different times with different meanings.

Such unpredictability could lead you to lose sight of the high degree of regularity of communication affairs, but a knowledge of the process of scientific inquiry, coupled with an understanding of communication science, can enhance your ability to gather information about communication processes in any discipline.

REFERENCES

Babbie, E. R. 1992. *The practice of social research.* 6th ed. Belmont, CA: Wadsworth.

Berger, C. R., and S. H. Chaffee. 1987. *Handbook of communication science*. Newbury Park, CA: Sage.

Dowling, C. L., and T. Lehnert, eds. 1992. List of journals indexed in AGRICOLOA 1992. Beltsville, MD: USDA, National Agricultural Library.

Elsevier Science Publishers. 1989. EMBASE. Amsterdam: Elsevier.

Engineering Information, Inc. 1990. *Publications indexed for engineering*. NY: Engineering Information.

Kerlinger, F. N. 1986. *Foundations of behavioral research*. 3rd ed. New York: Holt, Rinehart and Winston.

National Library of Medicine. 1993. List of journals indexed in *Index Medicus*. Washington, DC: National Library of Medicine, U.S. Dept. of Health, Education and Welfare.

Peters, J. 1992. Book industry statistics from the R. R. Bowker Company. *Publishing Research Quarterly* 8(2): 12–23.

Schramm, W. 1963. *The science of human communication*. New York: Basic Book.

CHAPTER

2

Extracting Information with a Problem-Solving Methodology

Logic in research is the link between the scientific methodology and a problem-solving methodology. If the purpose of research is to gain knowledge and if knowledge is organized information that explains and predicts, then organizing information into a useful body of knowledge should rely upon logic. Logic could be defined as the science that systematically directs mental operations toward attainment of a truth. Since problem solving is the act of applying logic to a problem, then problem solving is the logical thread woven through the fabric of a scientific methodology.

THE PROBLEM-SOLVING METHODOLOGY

For more than two decades the University of Wisconsin's Space Science Research Center tackled a variety of technical and scientific problems using a problem-solving methodology. Researchers Verner Suomi and Thomas Haig used a seven-step problem-solving method (fig. 2.1).

While steps 1 and 3 of the Wisconsin strategy are similar to the scientific methodology's approach to articulating problem statements (step 1) and data collection, analysis, and interpretation (step 3), steps 2, 4, 5, and 6 are unique. According to Suomi and Haig, when you identify solution criteria (step 2), you develop a framework for generating potential solutions (step 4), comparing solution criteria with potential solutions (step 5), and selecting viable solutions (step 6). These steps augment a scientific methodology by explicitly incorporating logic, or a problem-solving element.

Consider each of these logical enhancements in detail.

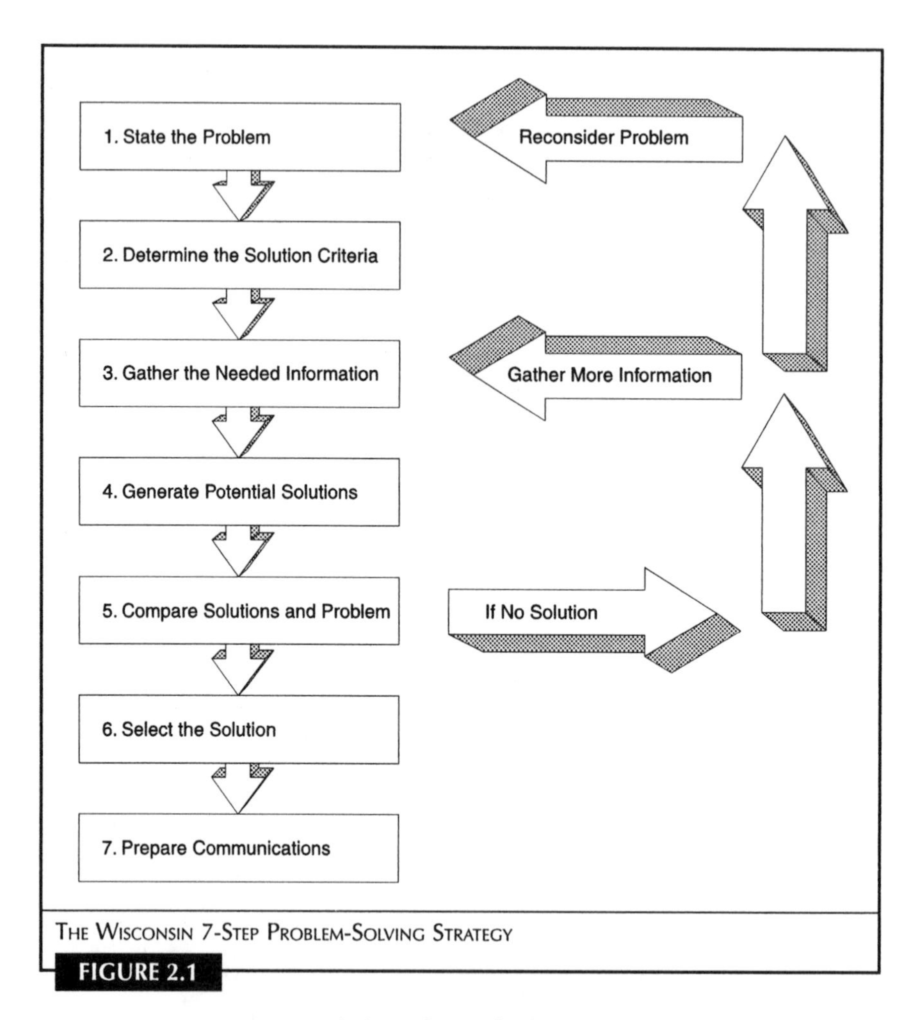

THE WISCONSIN 7-STEP PROBLEM-SOLVING STRATEGY

FIGURE 2.1

Determining the Solution Criteria

Solving any problem, like our wellness program example from chapter 1, requires a variety of resources including 1) skilled and knowledgeable staff, 2) available time to solve the problem, 3) available money for the problem solving activities, and 4) ethical considerations. Lack of available resources may severely restrict the potential solutions to a problem.

In *Conceptual Block Busting*, James Adams (1976) argues that problem solving requires fluency in the language or symbolism of a particular problem. Thus, a person without advanced mathematical skills cannot solve engineering problems, just as a person with no musical ability cannot tune a piano. Solving computer program problems, as you may know, often requires both the knowledge of the particular programming language and skills in writing software programs.

Time and money often limit viable solutions. While some problems might take three years to solve, you may have only three months. Clearly identifying the time limitations will help in the decision-making process under step 5. The available finances must also be considered. As an example, consider how your choice of a new car would be influenced by the available money. Just so, the available money influences how scientists, engineers, and technical specialists solve the problems they face and the potential solutions they consider.

Finally, ethical, moral, or legal considerations may occasionally rule out potential solutions that might harm individuals, animals, the environment, or society. The vast majority of scientists, engineers, and technical specialists carefully consider the ethical, moral, and legal criteria when considering solutions; universities, colleges, and research organizations also have committees that review a wide range of research and research-related areas: human research, animal care and use, biosafety, drugs and other controlled substances, and radioactive substances.

Generating Potential Solutions

As you gather salient information about a problem, a series of potential solutions begins to emerge. Some may prove quite useful but others will not. When trying to solve a particular problem, develop three or more alternative solutions. As you work, do not eliminate potential solutions. Eliminating solutions too quickly may create problems later. Keep in mind that you will soon be comparing the solutions against the criteria established in step 2.

Comparing Solution Criteria with Potential Solutions

One way to compare the solution criteria with the potential solutions is to build a table that has the criteria listed in rows and the solutions in columns, as shown in table 2.1. Such an arrangement allows you to compare different criteria against the solutions. Consider each criteria individually as you assess which solution might be the most effective. Think about which criteria are more important to assess the potential solutions.

Selecting the Solution(s)

Oftentimes, a clear winner emerges from the potential solutions. At other times, when some solution criteria are unmet, ask whether the solution can be adapted, modified, or changed to overcome its limitations. For example, a company might find it too expensive to provide a full wellness program, but may find that its employees would be willing to pay a modest fee to attend specific workshops. Flexibility and adaptability enhance solutions.

TABLE 2.1

BUILDING A COMPARATIVE PROBLEM-SOLVING TABLE

A Model Table for Comparing the Criteria and Potential Solutions

	Potential Solutions		
Criteria	A	B	C
One			
Two			
Three			

Even if no clear solution emerges, perhaps the problem-solving approach helped you more fully understand the problem. If so, perhaps you can now restate the problem, view it in another way, or reconsider your approach. In such cases, you may find it necessary to move through the problem-solving process again, as suggested in the Wisconsin strategy.

Preparing the Communications

After you have collected, analyzed, and interpreted your data using the scientific methodology, you will need to report your solution to the problem. You may need to make periodic reports throughout the problem solving process or simply write final reports and provide oral presentations.

APPLYING PROBLEM SOLVING TO INFORMATION GATHERING

Using a problem-solving strategy for identifying your information needs centers around

- Writing out your questions
- Determining what you know about the topic
- Determining how you know about the topic
- Identifying key terms, concepts, and alternative terms
- Identifying knowledgeable experts

Writing Out Your Questions

The key to finding information revolves around writing down the question as well as the subquestions that may help you identify the problem. The more narrowly you define the problem, the more clearly you can identify your information needs. You may start with a broad question, but most professional problem solving focuses on narrowly defined questions.

Developing a narrowly defined question is not easy and often requires repeated attempts to articulate your question. For example, medical researchers, epidemiologists, and health officials faced a problem of identifying the cause of the Four Corners flu, which in spring 1993, killed at least 16 people in the Four Corners area of Utah, Arizona, New Mexico, and Colorado (Scanlon 1993). Initially, medical investigators asked general questions such as

- What caused these deaths?
- What were the symptoms?

As the investigation progressed, the questions moved to greater specificity, such as

- What virus is causing the disease?
- How was it transmitted?
- How were the victims being infected?
- Were there other cases that had gone undetected? If so, for how long?
- What treatment may prevent deaths?

Eventually, medical researchers learned the problem was a hantavirus transmitted by the white-footed mouse through its feces. Researchers and epidemiologists were then faced with new questions such as

- Can we prevent the disease? If so, how?
- What precautions should people take to avoid the white-footed mouse?
- What precautions should they follow to prevent contracting the disease?
- What's the history of the disease?
- When was it first identified?

With these questions, medical researchers could review *definite* symptoms, conduct *specific* tests to determine the cause(s), and review *certain* medical literature to understand the epidemiology of the disease and its history, and to determine which, if any, treatments proved effective.

Consider another example. Some computers, monitors, printers, electric blankets, televisions, radios, other electric appliances, and electrical power lines generate low-level radiation fields. A national debate is emerging concerning the potential dangers of low-level radiation to people's health. Assume you are being assigned a project that will investigate this potential impact. While a general question such as "What impact does low-level radiation have on the public health?" can guide an initial literature review, it may be too general for scientific investigations. More specific questions needed to investigate any possible relationships might include

- Is there an increased level of leukemia among people living within 100 yards of major electrical power transmission lines?
- Is there a higher rate of spontaneous abortions among women working at video display terminals for more than four a hours a day?
- Is there a higher rate of other disease among people who live within 100 yards of major electrical transmission lines?
- Are there higher rates of leukemia among people who use electric blankets?
- Are there higher rates of leukemia among people who use electric blankets as their hours of use per year increase?

Scientific research requires specific, carefully conceptualized questions. Issues to be considered would include the type of power lines, the level of power being transmitted, the length of time people spend daily within the potential radiation field, the number of years they've lived close to the power lines, the distance from the power lines, and so forth. Detailed research questions that carefully explore the relations are needed to rule out alternative explanations.

Consider now a technical problem. At a nuclear power plant, a shaft connecting a motor to a pump circulating coolant broke down every three months. While a general question might be "What factors contributed to the breakdown of the shaft?" more specific questions might explore

- When did the shaft break down?
- Who manufactured the shafts?
- What materials were the shafts made of?
- Was the breakdown the result of metal fatigue?
- Was the breakdown the result of improper lubrication?
- Was the motor mounted securely on its mounts?
- What level of vibration did the motor generate? And was the level of vibrations within its allowable tolerance?
- Was the pump vibrating? And was that vibration within the allowable tolerances specified by the manufacturer?

- Was the shaft mounted securely with adequate bearings?
- What was the maintenance schedule for the motor? Shaft? Pump?
- Was the maintenance being performed? And was it performed correctly?

These questions illustrate the range of potential problems to be explored to solve the problem. Detailed engineering specifications for the equipment would help guide the subsequent problem-solving investigation.

Once you generate the questions to direct the problem solving, you can move to identifying terminology for information gathering.

Determining What You Know

Once you focus on a particular problem, write down what you already know about it, how you know what you know, and what you need to know. Try brainstorming. Turn off your internal editor, or critic, and write as fast you can about the topic. Don't worry about being correct or including details. Try writing lists or paragraphs, and start a new one every time you have a new thought on your topic. After you generate the list or paragraphs, ask, "How do I know? Where did I learn about the topic?"

For example, suppose you work for a small manufacturing company that builds widgets. Your boss tells you that the company has had an increasing number of workers' compensation claims from repetitive stress injuries and that he wants you to investigate the problem and determine if the company can reduce these claims. You have observed people assembling the widgets, and you realize that they do the same task over and over again with the same arm and wrist motion. You think that the problem is a lot like carpel tunnel syndrome and tendinitis that technical writers, computer programmers, and other computer users sometimes have, and so you list what you know about carpel tunnel syndrome and tendinitis.

- Computer users have trouble
- Slow, mild injuries occur over time
- Repetitive actions may cause the problem
- Some people wear wrist braces to solve these problems
- Some computer uses have adjustable chairs and tables
- People use wrist pads
- Changing the work habits helps overcome the problem
- Worker's compensation claims have cost thousands of dollars

For many topics, you will be surprised about your general knowledge of the topic. Keep in mind that what you know about the topic when you start a search colors, or influences, how you will see the topic. So always ask yourself, "How do I know what I know?"

Determining How You Know

Think about your experiences on the topic and then generate a list of the sources of your information. Keep in mind that people glean information from a wide range of sources: colleagues, seminars, journals, magazines, company reports, newspaper articles, newsletters, conferences, conference proceedings, and more.

Considering the repetitive stress syndrome problem, sources of information might include

- Articles in the local newspapers
- National news magazines
- A friend talking about personal repetitive stress injuries
- Classroom presentations while in college
- Newsletter articles
- Television news stories on the topic

If you can recall when the articles appeared in the various publications, retrieve the articles. Keep alert in your current readings and conversations for current accounts or reports on your topic. In some cases, you will be surprised at how frequently the general and specialized media cover a topic. Reviewing such articles and accounts also helps you identify concepts, terms, and variables under which you might search for more related information.

Identifying Key Concepts and Terms

To conduct an effective information search, ask yourself two questions:

- What major topic area(s) might the subject be listed under?
- What terms, concepts, or variables should you use in your search?

When trying to determine the major area(s), think about the broad-ranging subjects under which the topic might be classified or cataloged. Consider which professionals might concentrate on solving that kind of problem. For example, the question of the potential dangers of low-level radiation appears to fall under the general categories of medicine, health science, and possibly epidemiology, while the question of the flexible shaft breaking down appears to fall under the categories of engineering, mechanical engineering, and nuclear engineering.

Now you are ready to generate a list of key terms to guide your literature search and interviews. Start with a list of general terms. To illustrate, search terms for the question on low-level radiation and potential health problems might include

> radiation, low-level radiation, health, health risks, cancer, abortion, spontaneous abortion, leukemia, power line transmissions, computer terminals, computers, ergonomics, cataracts, high-level radiation

Potential search terms for the problem of the flexible shaft failure might include

> shafts, flexible shafts, failure, maintenance, lubrication, design criteria, stress, vibrations, engineering design, engineering specifications

Once you have a working list, talk about your problem with fellow students, colleagues, and subject matter librarians who have expertise in specialized areas. Ask what other subject areas and terms might be useful in searching the literature on your topic. As you search the literature, and find your first articles on the topic, look for related terms to help you identify other relevant literature. Add them to your growing list of terms to help direct your literature reviewing.

Identifying Knowledgeable Experts

Experts can be extremely helpful in facilitating your understanding of the problem. As you begin to think about a problem, consider what kind of professional might be knowledgeable on the topic and where you might contact them.

Large companies, government agencies, and colleges and universities often have a wide range of experts among their staff members. Even if such organizations do not have an expert on your specific topic, they may have an expert in the general subject of your topic.

For example, universities, government agencies, and larger companies often have an environmental hazards officer, nurse, physical therapist, or possibly a medical doctor, all of whom may have experience working with repetitive stress injuries. Once you identify potentially knowledgeable people, keep in mind that you should background yourself on the topic and prepare for an interview before visiting with such individuals. (Interviewing individuals will be discussed in chapter 7.)

REFERENCES

Adams, J. 1976. *Conceptual blockbusting*. San Francisco: San Francisco Book Co.

Scanlon, B. 1993. "Detectives" follow the trail of the Four Corners flu. *Rocky Mountain News*, 21 June 1993, 5A–6A.

CHAPTER

Identifying Library Resources

Most communities have several libraries, including university and college libraries; city, county, and state public libraries; local grade school, junior high, and high school libraries; government agency libraries; and business and industry libraries. Large universities generally have several libraries. The University of Wisconsin-Madison (UW), for example, has more than five million books, journals, magazines, and periodicals in its 24 major campus libraries and more than 50 smaller libraries, which specialize in a wide range of scientific and technical subjects: agriculture, astronomy, biology, chemistry, clinical sciences, engineering, geology, health sciences, life sciences, limnology, math, pharmacy, physics, and plant pathology.

While different libraries in a community may have the same publications, periodicals, and material, duplication of resources is often minimized to save money. Thus, a university research library will provide journals appropriate for its scientists while the city library is more likely to provide magazines for the practitioner. For example, when investigating radon, Ed Carpenter, a homeowner, discovered that Colorado State University, with its 1.6+ million volumes, did not have trade magazines written for home builders and contractors. The local library, on the other hand, did have the business trade magazines he needed.

Whether you use a university library, a government library, or private library, to learn about the library so that you can use it efficiently

- Obtain publications explaining the library
- Attend orientation programs, workshops, and seminars
- Discover the location of major resources
- Learn the cataloging system

OBTAINING PUBLICATIONS EXPLAINING THE LIBRARY

When you first walk into any library, look for any literature, bulletin boards, kiosks, audiovisual aids, posters, floor diagrams, or other orientation materials. Libraries usually place such materials near the main entrance. Such literature may provide a floor plan of the library as well as library use guides such as catalogs, call number or cataloging systems, electronic databases, and computer guides. Some libraries also provide bibliographies and lists of useful resources by disciplines.

When familiarizing yourself with a library, collect the general materials that will help you use the library as well as the specific materials covering your respective discipline. Considering creating a library notebook and including these materials in your notebook. As you familiarize yourself with different library resources, add notes to your library notebook.

ATTENDING ORIENTATION PROGRAMS, WORKSHOPS, AND SEMINARS

Libraries often offer courses, workshops, and seminars to help users become more proficient in using the library. Some universities and colleges provide a one- to three-credit course that provides an orientation to the institutions library system and its services. Some libraries offer advanced library courses to familiarize users with the literature of academic disciplines such as social sciences, humanities, biological sciences, or engineering. Others have multimedia presentations or pamphlets that provide a self-guided tour of the library's resources. Many libraries also offer special electronic workshops that provide an introduction to online catalogs, CD-ROM technologies, and software to make library use more efficient.

Libraries often post announcements of upcoming seminars and workshops near their main entrances. Such seminars and workshops may also be announced through the institution's newsletter, student newspaper, memos to faculty and staff, and electronic mail when available.

DISCOVERING THE LOCATION OF MAJOR RESOURCES

Once you know the library's holdings and its cataloging system(s), learn the physical layout of the library. Libraries usually have major areas, floors, or rooms including a general reference area, stacks, periodicals room, government publications, checkout desk, staff offices, and interlibrary loan services. Keep in mind that each library usually has a different organizational layout of these areas.

General Reference Area

The general reference area(s) may include the general reference publications, the abstracting journals and indexes, card catalogs, computer terminals for accessing the online catalog, and a CD-ROM collection.

Each library's physical arrangement varies. In some cases the libraries will put all reference areas together on one floor or in one location; other libraries will have one area for general references, a nearby room of computer terminals to access the online card catalog, and a separate area for the abstracting journals and indexes. Some libraries place general references in one area, scientific and technical abstracts in another area, the social science abstracts and indexes in another area, and the humanities and education abstracts and indexes in still another area.

Stacks

The bound volumes, books and periodicals alike, are located in the stacks, which are the bookshelves containing the library's collection. Libraries organize most of their collection in stacks according to the call number system used. Signs clearly mark the stacks and identify the areas by call numbers. Library leaflets usually indicate which floor and stack hold particular call numbers. Some online catalogs provide both the call number of a volume and its location; in other cases, the system will direct you to the library's current holdings catalog or printout.

Most libraries have open stacks so that you can obtain the materials you need on your own. A few have closed stacks with limited general access. In some libraries, the older publications are moved into a storage area of closed stacks. No standard guidelines dictate which items go into storage and which stay in the open collection. Thus, for one library, publications older than 30 years may be placed in storage, while at another library items must be over 40 years old. To obtain materials from closed stacks you must request them. Library staff will retrieve the materials for you to pick up at a specified desk.

Periodicals Room

Many libraries have a periodicals room that contains current journals, magazines, and other recurring publications. Some periodicals rooms contain only current publications (the last two years) of the quarterly, monthly, or weekly periodicals on shelves. After two years, the library sends the periodicals to be bound by year or volume number. These bound periodicals are then shelved in the stacks. Other libraries' periodicals rooms provide the current periodicals on shelves and the bound volumes in stacks in the same room. Such libraries may limit the volumes in the stacks to the more frequently used periodicals.

Government Publications Stacks

Many libraries have a collection of government publications that are kept in the government documents stacks using the Superintendent of Documents cataloging system. Some federal publications are published in bound hardcover volumes; others are released as 8.5"x11" leaflets, pamphlets, posters, bound reports, or in a variety of other formats, including electronic.

Some libraries house state and local government publications in one area, while other libraries integrate state and local government publications into the library's overall collection. A careful review of the library's literature or catalog system, or asking a librarian, should reveal the library's current system.

Checkout Desk

The checkout desk is usually located near the main doors, and its staff can provide help in obtaining volumes. Check-out staff members may be part-time students or clerical employees who can answer general questions but may lack the expertise of full-time subject matter and reference librarians.

Checkout policies vary among libraries. Check your status as a borrower. For example, if you are not a student or faculty member, you may not be able to borrow materials at a college or university library. Some libraries do not check out particular publications such as major reference works and current periodicals. Libraries often limit the checkout period for bound periodicals to two to three days. At college or university libraries, the publication checkout period often runs two to three weeks for students but may run the academic term for faculty and staff.

Checkout procedures vary among libraries. Most libraries now use an electronic checkout system. Most books have a machine-readable bar code. The checkout clerk scans the user's university or college identification or library card and then scans the book's bar code. Such computerized systems speed the checkout system and simplify circulation management. Some libraries are now implementing self-checkout stations that are simple to use and have directions attached. A few libraries have not added bar codes to all of the volumes and may not have earlier books entered into the online catalog. In such cases, you may need to fill out a checkout card, or the checkout clerk may need to enter the key volume information into the library's checkout system.

Many publications have "tags" that can trip an alarm should you try to leave the library without checking out the publication. The clerk who checks out the publication desensitizes the "tag" so that it usually will not trigger the alarm. Many libraries have turnstiles and scanner devices that you must walk through when exiting the library. At this point, if you trip the alarm, you must return to the checkout desk to have the publication checked out and/or properly desensitized.

As you become familiar with the library, learn its checkout policies and procedures, the return policy, the notification policy used to remind you of checked out materials, and the fines charged for late returns.

Library Staff and Offices

Research libraries often have subject matter specialists—librarians who are especially skilled and knowledgeable in specific disciplines. For example, a library might have a life sciences librarian, a physical sciences librarian, a social sciences librarian, and a humanities librarian. Each one develops an in-depth knowledge of the literature in the fields under their assignment. They have one or both of the following responsibilities: 1) building the library's collection in their area of specialization, and 2) working closely with those who need help searching for specific information in disciplines within their broad field.

Such librarians also have expertise in one or more of the specific topics within their area. For example, a physical science librarian may also be an expert in the chemistry literature, a social science librarian may be an expert in the psychology literature, or a biological sciences librarian may be an expert in the agriculture literature. Frequently, such librarians have gained some of their expertise by having earned an undergraduate or masters degree in one of the specific fields before completing a library science degree. Such specialists are knowledgeable on a wide range of publications and can be of great help to you in searching the literature in specific fields. They are also skilled in using the abstracts and indexes as well as CD-ROM bibliographic databases and the available databases for their respective fields. They can help you plan and answer questions that develop as you search for specific publications and articles.

Interlibrary Loan Services

Since libraries cannot purchase all of the publications you may need, most libraries have loan arrangements with other libraries to obtain journal articles, books, technical reports, theses and dissertations, and other materials. If your library does not have a publication, you may be able to request the materials through the library's Interlibrary Loan services (ILL). To do so, you fill out an ILL request. The ILL staff will try to locate the item you need at another library willing to loan the item.

Libraries will usually loan books and full technical reports, but will photocopy articles out of journals, magazines, and other periodicals. Some libraries charge you for the item and other libraries do not. Costs, if any, vary with the institution. The required time for ILL also varies from a week to six months, depending upon the library making the request and the library holding the needed materials.

In a few cases libraries cannot obtain publications through interlibrary loans. Other libraries limit the numbers of requests you can make at any one time. Some libraries do not loan theses and dissertations, but others do. Obscure publications sometimes are not held by any other libraries, and specialized items such as specifications and highly expensive publications are not loaned.

LEARNING THE CATALOGING SYSTEM

With tens of thousands of titles, research libraries usually catalog their holdings using either the Library of Congress classification system or the Dewey decimal system, and libraries catalog federal government publications using the U.S. Superintendent of Documents cataloging system.

The Library of Congress classification system uses 21 categories, employing the following letter designation system.

Letter	Subject
A	General works
B	Philosophy and religion
C	Auxiliary sciences of history
D	History (except American)
E-F	History of America
G	Geography, anthropology, folklore
H	Social sciences
J	Political science
K	Law
L	Education
M	Music
N	Fine arts
P	Language and literature
Q	Science
R	Medicine
S	Agriculture
T	Technology
U	Military science
V	Naval science
Z	Bibliography and library science

The Dewey decimal system uses a number system for its categories.

Number	Subject
000	General works
100	Philosophy
200	Religion
300	Social sciences
400	Language
500	Pure science
600	Technology (applied sciences)
700	Arts
800	Literature
900	General geography and history (Katz 1979)

A review of the two systems shows that the Library of Congress system provides more categories and divisions covering scientific and technical fields than does the Dewey decimal system. University and college libraries usually use the Library of Congress cataloging systems; high school, junior high, grade school, and public libraries often use the Dewey decimal system.

U.S. federal government publications use the Superintendent of Documents (SuDoc) classification system. Adopted in the early 1900s, the SuDoc call numbers are the system whereby the federal government classifies its documents. SuDoc call numbers consist of a letter or letters and a series of numbers. The letter or letters identify the U.S. government department from which the publication comes and the numbers identify the agency or publishing unit. Colorado State University's Morgan Library provides this example:

A1.36.21DOC

A	Department of Agriculture
1	Secretary's Office
.36	Technical Bulletin Series
.21	Volume number within the series
DOC	Location symbol—where it's located in the library

While the SuDoc system itself is straightforward, searching is often difficult because federal agencies tend to shift the responsibilities for different scientific and technical subjects between agencies, and several agencies may publish on the same scientific or technical topic.

KEEPING UP WITH THE CHANGING LIBRARY

While many of the main resources remain the same, the move to electronic systems is rapidly changing many libraries. Major research universities, small colleges, government agencies, and businesses now link their internal networks to even larger, perhaps worldwide, information networks. To keep abreast of the changes, visit your library regularly, attend workshops and seminars on how to use the library, and ask for the latest updates and handouts explaining library resources and how to use them. If you have a problem, do not be afraid to ask the library staff for help. The vast majority of staff members are eager and willing to help you.

Finally, keep in mind that however libraries change, having a strong foundation in understanding and using both the printed and electronic resources will ensure your success in searching for information.

REFERENCE

Katz, W. 1979. *Your library*. New York: Holt, Rinehart & Winston.

CHAPTER

Using Printed Resources

Printed resource searching skills are invaluable. Keen information gathering from printed resources entails a focused approach that includes developing a literature review problem statement and devising a strategy to search printed resources.

While the following discussion provides a framework to build your literature searching skills, keep in mind that a systematic approach, as illustrated in figure 4.1, saves time and improves your chances of identifying which literature will help solve your particular problem.

DEVELOPING A LITERATURE REVIEW PROBLEM STATEMENT

An effective literature review begins with a well-developed problem statement that identifies the key concepts and terminology under investigation.

To illustrate, assume that you work for a small company that desires to cut personnel health insurance costs by implementing a health-fitness program. This company is aware that health programs have successfully lowered health insurance costs for other companies, has already implemented smoking cessation and drinking programs, and is now considering a weight control program. The company asks you to investigate the success rate of currently existing weight-reduction programs. If you approach this investigation incrementally, you can develop a salient research question that increases the likelihood of a successful literature review.

First, you talk with the nutritionist at the local hospital and learn that the American Heart Association has a Slim for Life Program that provides guidance on changing eating practices, reducing fat consumption, and increasing aerobic exercise. This program concentrates on reducing an individual's cholesterol by reducing the amount and kinds of fat consumed.

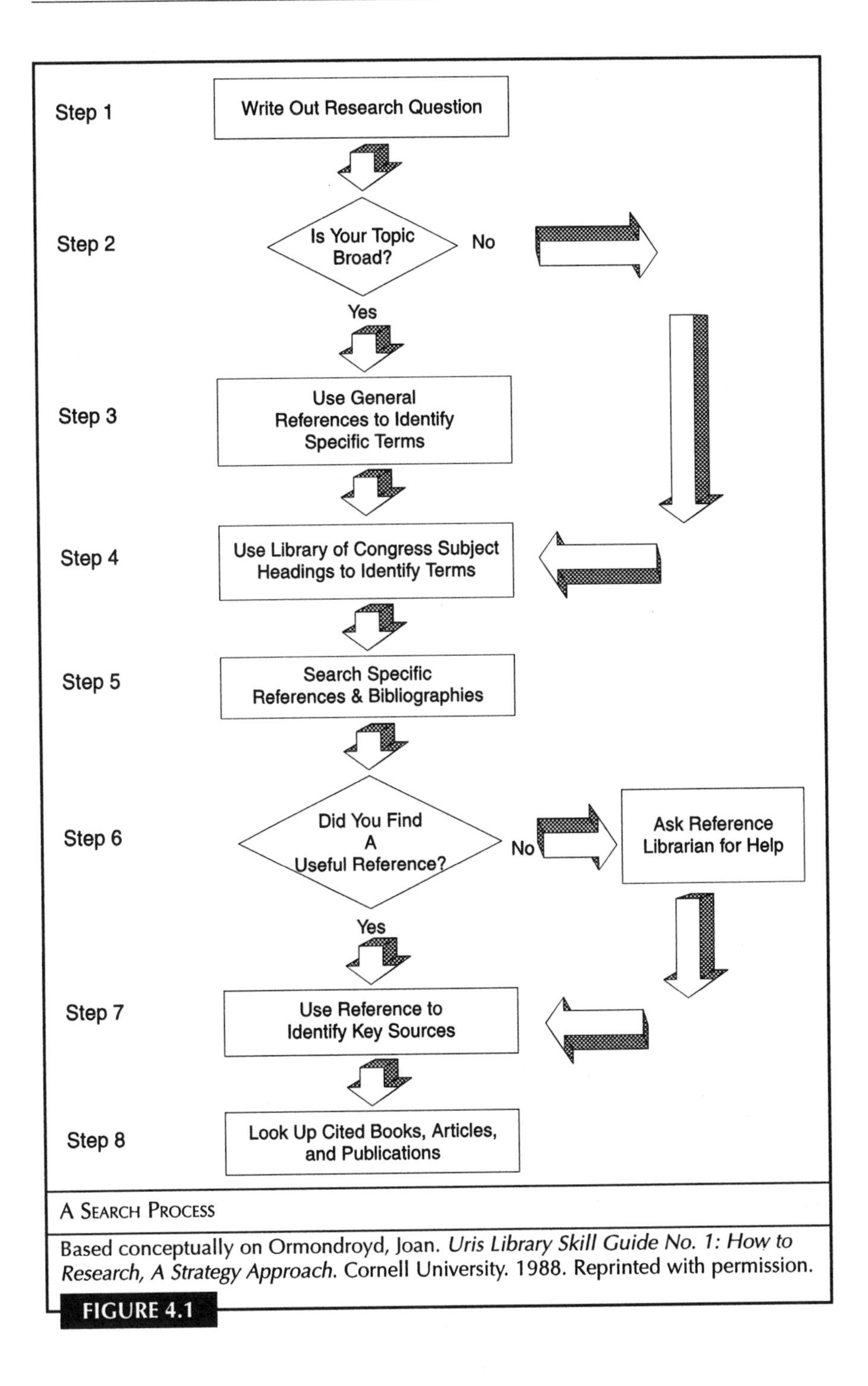

A SEARCH PROCESS

Based conceptually on Ormondroyd, Joan. *Uris Library Skill Guide No. 1: How to Research, A Strategy Approach*. Cornell University. 1988. Reprinted with permission.

FIGURE 4.1

Second, because you must ascertain whether such programs work, you ask, "Does the Slim for Life Program improve employees' health?" Such a broad question, however, may not elicit the specificity you need. What does "weight control" mean? What does "improve" mean? What does "employee's health" mean? Does "weight control" mean maintaining a current weight? Does "weight control" mean reducing to an "ideal weight"? What constitutes an "ideal weight"?

Third, to narrow the scope of questioning, you query just one aspect of the Slim for Life Program—increasing aerobic exercise. You now ask, "Does aerobic exercise help reduce cholesterol?" From this narrowly postulated problem statement, you identify key concepts—aerobic exercise, reduce, and cholesterol.

Next, generate a list of terms that define your key concepts. For the concept of aerobic exercise, you may include terms such as aerobic exercise, swimming, jogging, bicycling, horseback riding, and walking. For the concept of cholesterol, terms may include cholesterol, high density lipids, low density lipids, and lipids. For the concept of reduce, terms may include reduce, lower, and minimize. You are now ready to begin your literature review.

As in all problem solving, the more narrowly you can define the problem statement, the more quickly you can solve the problem. Seldom will broad, general statements provide the specific terminology and thereby the guidance you need to find information for a specific problem.

DEVISING A SEARCH STRATEGY

After you identify subject-specific terminology, you need a systematic strategy for researching those terms. A methodical strategy includes looking at general information sources, exploring specific abstracting publications and services, and retrieving needed books, reports, articles, and publications.

You may choose to search for these sources through the printed matter we will discuss in this chapter, or you may choose to search electronically, which will be discussed in chapter 5.

While specific library holdings differ, most libraries have the following resources/tools to help you identify and locate printed material.

- Card catalog
- Reference room
- Printed abstracts and indexes
- Periodicals
- Books and monographs
- Government publications

- Maps, special collections, and archives
- Microforms

PUBLIC CATALOG

While the card catalog has historically been the primary resource for identifying a library's collections of books, periodicals, and other materials, libraries are quickly converting these catalogs into electronic formats that will be discussed in chapter 5. Even if you find that you prefer using electronic resources, skill in using printed sources could be valuable in a variety of situations where computer systems are not available. Alternately, as many libraries convert to electronic online catalogs, some libraries may remove card catalogs completely or relegate them to the basement or storage. Whether you use a card catalog or an online catalog, you should ask

- Does the catalog contain all of the library's holdings?
- Does the library have more than one catalog?
- Does the catalog contain entries for all books, periodicals, and government publications?
- What are the limitations of the catalog?

Card catalogs consist of large cabinets that contain drawers of 3″x5″ cards. Each card will be either a subject, title, or author card, containing distinctive information. The author card provides details on each publication, beginning with the call number and author's name and including the title, place of publication, publisher, physical description of the publication, subject headings, and library cataloging information. The title card contains similar information, but begins with the call number and publication title. The subject card begins with the call number and the subject followed by the other information. Keep in mind too that some libraries may interfile the cards or combinations of the three cards.

Begin searching the card catalog under the terms you identified. Note the titles and authors of relevant publications as well as additional terms under which you might search. Look up the titles and authors as needed. Some authors write multiple books on related subjects, so check authors' names for related publications.

Once you identify a source, take thorough notes (panel 4.1). Begin by searching the subject volumes of *Books in Print* against your list of terms. Once you find a publication and an author, search the author volumes for related works by the same author. If you find a book written by two authors, search under both authors' names.

If you do not find any books on your topic, you may have defined your topic too narrowly. If so, write a broader problem statement, identify key con-

PANEL 4.1

Preparing Citation/Source Note Card

If you're using a card catalog, make sure you take notes on the following.

- Full name of the author
- Date of publication
- Publisher
- City of publication
- Key terms on the bibliographic card or electronic citation
- Call number
- Location in the library, if available
- Where you found the citation (original source and page number)

The last entry is critical should you encounter problems finding the particular citation. With this information you can backtrack, check the citation again, and find it. In taking notes, everyone errs on occasion, and providing the final entry simplifies making the needed corrections.

If you're using an online catalog or electronic bibliographic databases, make sure that the printout or screen capture (downloading the information on the screen to a disk) includes the key information. If it doesn't, retrieve that information and add it to your notes.

Here's an example of a typical note card:

Gannon, Robert

1991

Best Science Writing

Phoenix, AZ: Oryx Press 1991

Terms: Science writing, science journalism, science communication

Call number: T11 B46 1991

Location: Morgan Library, 4th Floor East Stacks

Gannon selected 12 science articles that had won national awards beginning in 1973 through 1988. He introduces each article with the author, award won, where it appeared, and the award received. Gannon then provides a review of the article and discusses the problems the author faced in writing the article.

Source: Online search on CSU library catalog.

cepts and terminology, and repeat your search. Once you learn specific titles or authors, you can search the library's card catalog or online catalog. If the library does not have the recent title, request it through interlibrary loan.

REFERENCE ROOM

The reference area or room often includes almanacs, encyclopedias, dictionaries, style guides, statistical summaries, who's who and biographical publications, guides to libraries, usage guides, and other general reference materials. Publications of special use when you are exploring a field include

- *Books in Print*
- *Gale Directory of Databases*
- *CD-ROMs in Print*
- *Gale Directory of Publications and Broadcast Media*
- *Literary Market Place*
- *Reference Sources in Science, Engineering, Medicine, and Agriculture*
- *Information Sources in Science and Technology*
- *Guide to Reference Books*
- *Walford's Guide to Reference Material*
- *Standard Rate and Data*
- *Ulrich's International Periodicals Directory*
- *Library of Congress Subject Headings*

A check of *Books in Print* will enable you to identify the most recently published books on your topic. Before you begin, spend a few minutes familiarizing yourself with its different volumes, organization, and information provided. *Books in Print* provides an annual listing of current books published in the United States as well as a listing of books soon to be published. *Books in Print* comes in printed volumes as well as CD-ROM format. The printed volumes are arranged by title, subject, and author, and you can search the CD-ROM by the same categories. In either format, *Books in Print* is a useful tool for identifying current books covering a specific topic. *Books in Print* also provides the names and addresses of all publishers whose books are listed.

The *Gale Directory of Databases* identifies the rapidly expanding list of computer research tools. The 1993 Volume 1, *Online Databases*, identifies and details more than 5,200 publicly available databases. Volume 2, *CD-ROM, Diskette, Magnetic Tape, Handheld, and Batch Access Database Products*, identifies and details more than 3,200 databases. The first part of each volume describes the databases and their features; the second part identifies database producers with listings of their products; and the third part explains how to contact the database vendors and distributors and identifies the products and their uses. Each volume also provides geographic, subject, and master indexes. The *Gale Directory of Databases* is also available online.

CD-ROMs in Print serves as an international guide to CD-ROM, CD-I, CDTV, multimedia, and electronic book products. Published annually, this directory identified 1,400 titles in its 1991 edition, 2,900 titles in its 1992 edition, and 3,502 titles in its 1993 edition, with more titles likely to be added with each annual edition.

The *Gale Directory of Publications and Broadcast Media,* formerly *Ayer's Directory of Publications*, identifies journals, magazines, newspapers, radio stations, television stations, and cable systems and provides publication titles, publishers, editors, and addresses. Users can search by state and subjects. The *Literary Market Place* lists major book publishers and provides their names and addresses. It too is indexed by subject and provides a way of identifying key publishers of well-known books.

Reference Sources in Science, Engineering, Medicine, and Agriculture provides a reference guide to the literature in these four fields. As its title suggests, Hurt's *Information Sources in Science and Technology* also identifies a range of specific publications covering scientific and technical fields.

Sheehy's *Guide to Reference Books* identifies abstract journals, bibliographies, dictionaries, encyclopedias, guides, handbooks, indexes, bibliographies, and other useful general reference publications. If you are just beginning to search a topic, check Sheehy's *Guide* to identify the general references covering specific topics. *Walford's Guide to Reference Material* identifies a range of scientific and technical references.

Standard Rate and Data includes a listing of major specialized and trade magazines. While its primary function is to provide the rates these publishers charge for advertisements, *Standard Rate and Data* can be used to identify specialized magazines serving professionals in technical, scientific, and related areas. *Ulrich's International Periodicals Directory*, available in both hardcopy and online versions, covers some 140,000 serials (periodicals) worldwide under 966 subject headings. *Ulrich's* fifth edition describes more than 7,000 daily and weekly newspapers in the United States. This edition also identifies 3,838 serials available online and 880 available on CD-ROM.

The *Library of Congress Subject Headings* reflects the organizational strategy used for cataloging and thus organizing for the Library of Congress call number system. Checking the topics by subjects allows users to identify related terms and terminology.

The publications identified in this section provide a starting point for exploring the general references available for the scientific and technical subjects. Carefully checking a library's catalogs and electronic databases will identify publication call numbers and their location. When searching for specific publications in reference sections, remember to check for other titles shelved nearby. You may find references that will identify useful resources for your literature searching.

In addition, many libraries publish lists of general references covering specific scientific, technical, and specialized topics. The University of Colorado in Boulder, for example, provides a series of handouts identifying general references and searching strategies for kinesiology and physical education, gerontology and geriatrics, foreign language technical dictionaries, and guides to style and research. Idaho State University at Pocetello provides a series of general handouts on reference sources for botany, ecology, and plant agriculture; clinical pharmacy; zoology, ethology, and ecology; and physics and astronomy. The University of Maryland at College Park provides a series of general handouts on such topics as agriculture, animal science, aquatic biology, botany, and ecology; international economic, industrial, and trade statistics; and style manuals.

By reviewing the general references on the topic, you can develop a broad understanding of the subject and identify specific bibliographies, indexes, references, and online bibliographic databases to search. For taking content notes, see the guidelines in chapter 6 on retrieving and evaluating information.

PRINTED ABSTRACTS AND INDEXES

Printed abstracting and indexing journals have been the key tool for identifying recently published research articles on particular topics. (See chapter 5 for a discussion of their electronic counterparts.) Much like *Readers' Guide to Periodical Literature*, scientific and technical abstracting and indexing journals provide an efficient and effective tool. Here is a partial list illustrating the range of technical and scientific fields covered:

Air Pollution Abstracts
Bibliography of Agriculture
Biological Abstracts
Chemical Abstracts
Communication Abstracts
Dissertation Abstracts
Energy Abstracts
Engineering Abstracts
Environmental Periodicals Bibliography
Food and Technology Abstracts
Forestry Abstracts
Index to Dental Literature

Index Medicus
International Nursing Index
Monthly Catalog of U.S. Government Publications
Oceanic Abstracts
Physics Abstracts
Psychological Abstracts
Science Citation Index
Sociological Abstracts

Before you begin your search, see if the library has general information sheets, or guides, on how to use the journal that you are reviewing. If so, carefully review the guidelines to develop a general idea of how to use the publication. Some abstracting journals provide lengthy introductions on how to use the various services; others provide separate instruction manuals. A few provide short courses. Some abstracts have multiple organization strategies for flexible searching by subject and author to identify a specific abstract by number, which you then look up. Some abstracting journals provide a full citation and abstract (summary) while others only provide the specific citation.

Once you identify a potentially useful article, read the citation. If it still looks potentially useful, prepare a note card of the citation details as suggested in panel 4.1.

The *Science Citation Index* is particularly useful in that it covers diverse academic fields such as agricultural, biological, and environmental sciences; engineering, technology, and applied sciences; medical and life sciences; physical and chemical sciences; and behavioral sciences. It provides bibliographic information on each article, lists cited authors and works, allows searching by key words in article titles, and identifies each author's corporate or academic affiliation.

PERIODICALS

Periodicals include journals, magazines, newsletters, and newspapers.

Journals include academic and research periodicals that report developments in their particular area. Such journals number in the tens of thousands since most technical and scientific fields have at least one major journal, and often a dozen or more, covering a range of topics within the field.

Scientific and technical journals provide articles 1) reporting research findings, 2) summarizing the research literature on a narrowly defined topic, and 3) solving or designing solutions to problems. Occasionally articles stating and justifying a position may appear, but these are more common in the humanities fields.

Magazines include both trade and special interest periodicals. Trade magazines usually serve professionals in a particular field with a variety of articles that focus on solving specific problems, reporting developments in the field, and covering professional activities. Special interest magazines focus on a single special interest such as computers, foods, homemaking, popular science, or technical fields.

With the advent of desktop publishing systems in the mid-1980s, newsletters have proliferated. The *Oxbridge Directory of Newsletters* identifies 20,000 newsletters published in the United States and Canada. The *Oxbridge Directory* defines newsletters as publications that usually build their circulation primarily through editorial preciseness, i.e., the content. Newsletters, usually 4- to 12-page monthly or quarterly publications, cover a wide range of topics but are usually narrowly defined to serve the particular interests of their audience. Newsletters usually cover topics of short-term or immediate interest and might summarize the research literature in a particular field, provide how-to or problem-solution information for specialists in a particular area, or provide information of more general interest. Some newsletter subscriptions are relatively inexpensive while others run several hundred dollars a year. Libraries often limit their subscriptions to newsletters.

Libraries also subscribe to local, state, national, and international newspapers. Most libraries subscribe to a small collection of newspapers, retain them for a limited time, and then remove back issues, sometimes replacing them with microform copies when they become available.

BOOKS AND MONOGRAPHS

Books include textbooks, monographs, and trade books. Textbooks usually represent the author's attempt to introduce students to the basic concepts, ideas, and ways of thinking in a specific field or discipline. Monographs, in contrast, often focus on a particular research project; summarize an extensive literature review of a topic or special research project; or offer a historical explanation of a topic or issue. Trade books often provide guidance on how to solve particular problems or implement specific ideas, or discuss related professional activities.

Check subject-specific textbooks and textbooks in related courses for their citations of other books, research articles, and publications covering the specific topic. Keep in mind that most textbooks do not contain citations of the more recent studies, but they may provide citations to within 12 to 6 months of the text's publication date. Frequently the sources cited in textbooks represent classic works on a particular topic and may provide exceptionally thorough and detailed literature reviews.

GOVERNMENT PUBLICATIONS

University and college libraries usually have a collection of government publications from federal, state, and local government agencies. Most are usually from the United States Government, the world's largest publisher. Every major federal government agency publishes books, periodicals, special reports, technical reports, and, now, material in electronic formats, including CD-ROM. Major university research libraries are often designated as depositories for federal publications and often maintain a major collection of government publications with one or more librarians specializing in overseeing that collection. Many government publications cover scientific and technical fields. Because your topic may be covered by several government agencies, you may need to explore the publications of several government agencies.

For example, natural resource and environmental topics may turn up in the literature of several U.S. departments—Agriculture; Defense; Energy; Health, Education, and Welfare; and the Interior—as well as other agencies.

Some state agencies also publish scientific and technical journals, magazines, newsletters, and reports. A library's holdings will depend, in part, on the library and state agency distribution policies. Generally, a library has more publications of the state in which it is located than of other states.

MAPS, SPECIAL COLLECTIONS, AND ARCHIVES

When reviewing the library's general orientation materials, see what maps, special collections, and archives are provided. Map collections may include world, country, state, and local maps as well as the standard topographic maps. The specifics vary with the library and reflect the institution's research focus.

Some libraries have collections of the papers and personal documents of well-know professionals. These collections, which usually have been donated to the library upon the person's retirement or death, provide a wealth of detailed information about an individual's life and professional activities. Special collections may also include those of historical and/or research significance in particular fields, such as the Bailey Botanical Collection, the Vietnam Fiction Collection, or the Germans from Russia Collection at Colorado State University.

University and college libraries also archive information about the institution. Such information might include the annual budget, technical reports, planning documents, and more, depending upon the institution's policy on archiving its administrative publications.

MICROFORM PUBLICATIONS

While online and CD-ROM technologies have emerged as major new ways of packaging information, microform publications such as microfilm and microfiche are a mainstay of many libraries. Libraries usually have a major area devoted to a variety of microform publications. Rather than retain printed copies of newspapers and magazines, libraries often provide selected publications on microforms because they require less space. Besides periodicals, libraries may turn to microforms for retaining copies of conference proceedings, special reports, and other one-time technical and scientific reports.

Microforms are created by photographing the publication. To illustrate, the ERIC microfiches generally hold about 75 8.5"x11" pages on one 3"x5" microfiche. Microforms require special reading machines that enlarge so that users can easily read the materials. Libraries usually locate microforms in an area where microform readers are provided. Microform readers may work on only microfiche units or on microfilm and microfiche. The readers consist of a platform, enlarging lens, and light system, usually with posted use instructions on or near the machines. Libraries may also have a few machines that can photocopy the microform materials. Such photocopies are often of poorer quality than those from printed sources.

REFERENCES

Bowker, R. R. 1993. *Books in print*. New York: R. R. Bowker.

Bowker, R. R. 1993. *Literary market place*. New York: R. R. Bowker.

Bowker, R. R. 1993–94. *Ulrich's international periodical directory 1993 to 1994*. New York: R. R. Bowker.

Desmarais, N. 1993. *CD-ROMs in print, 1993*. Westport, CT: Mecker.

Hurt, C. D. 1988. *Information sources in science and technology*. Englewood, CO: Libraries Unlimited.

Information Retrieval Ltd. 1993. *Genetics abstracts*. London: Information Retrieval Ltd.

Institute for Scientific Information. 1993. *Science citation index*. Philadelphia: Institute for Scientific Information.

Library of Congress. 1992. *Library of Congress subject headings*. 15th ed. Washington, DC: Cataloging Distribution Service, Library of Congress.

Marcaccio, K. Y., ed. 1993. *Gale directory of databases*. Detroit: Gale Research.

Malinowsky, H. R. 1994. *Reference Sources in Science, Engineering, Medicine, and Agriculture*. Phoenix, AZ: Oryx.

Oxbridge Communications. 1993. *Oxbridge directory of newsletters, 1993*. New York: Oxbridge Communications.

Sheehy, E. P. 1986. *Guide to reference books*. Chicago: American Library Association.

Standard Rate and Data Services. 1993. *Business publication advertising rates and data*. Skokie, IL: Standard Rate and Data Services.

Standard Rate and Data Services. 1993. *Consumer magazine and farm publication rates and data*. Skokie, IL: Standard Rate and Data Services.

Standard Rate and Data Services. 1993. *SRDS newspaper circulation analysis*. Skokie, IL: Standard Rate and Data Services.

Troshynski-Thomas, K., and D. M. Burek, eds. 1994. *Gale directory of publications and broadcast media*. Detroit: Gale Research.

Walford, A. J. 1991. *Walford's guide to reference material*. London: Library Association.

CHAPTER

5

Using Electronic Resources

Different libraries provide varying levels of electronic resources. Some libraries maintain the leading edge of systems and technologies, while other libraries are just beginning the transition to electronic systems. Clearly, the more you understand about electronic library systems and how to use them, the more successful you will be in identifying and retrieving the subject-specific literature cites you need and the more time you will save when using these systems. With the dropping prices of CD-ROM mastering equipment, the availability of CD-ROM readers for personal computers, and the emergence of an information highway, more and more information is becoming available electronically.

This chapter provides

- A strategy for searching online catalogs, CD-ROMs, and electronic databases
- An overview of online catalogs and databases and their use
- An exploration of CD-ROMs and their use, including online books, manuals, and other publications
- A discussion of commercial online databases
- A review of the systems for accessing electronic information

A STRATEGY FOR SEARCHING ONLINE CATALOGS, CD-ROMS, AND ELECTRONIC DATABASES

Begin your search by stating your research questions. Underline the key concepts or terms in your research question, and then generate a list of synonyms (panel 5.1). If this is your initial search, generate as many similar terms as possible. To find related or similar terms, ask others who may be familiar with

PANEL 5.1

LIBRARY WORKSHEET

LIBRARY SEARCH WORKSHEET

1. **Write down your research questions and underline your key words or phrases.**
   for example: nutrition in cats

2. **Make a list of SYNONYMS for each key word concept you underlined.**
   Include singular/plural forms, spelling variations, different word endings, etc.

3. **Connect the synonyms with the Boolean operator "or."**
   "Or" BROADENS your search to find records in which one or more of your terms appear.

feline
cats
cat

for example: *Cats* or *Cats* or *Feline*

Concept 1 = Set

________ or ________ or ________
or ________ or ________ or ________

Concept 2 = Set and

________ or ________ or ________
or ________ or ________ or ________

Concept 3 = Set and

________ or ________ or ________
or ________ or ________ or ________

4. **Connect the concept sets with the Boolean operator "and."**
   "And" NARROWS your search to require at least one term from each concept set to be present in each record.

4 5

______ and ______ and ______
(concept 1 set) and (concept 2 set) and (concept 3 set)

for example: 4 and 5
(cat set) and (nutrition set)

the terms; check dictionaries, textbooks, thesauri, and references; and, if you're having problems, check with a reference librarian or the library's specialist on the field. Keep in mind too that your search will turn up related terms that you can use.

As you develop your search, also develop a strategy to identify the best available sources in the least time once you are in the system and searching. Key to such searches is using a Boolean Logic search strategy, which specifies the relationships between terms by using the connectors "or," "and," "but not," or "and not." These different connectors tell the system how to narrow or expand your search. "And," "but not," and "and not" narrow your search, and "or" expands your search. Some systems also allow truncation—typing only part of the word and then adding a truncation symbol such as an asterisk (*) or pound sign (#).

Assume your research question is, "What does the research tell us about the effectiveness of multimedia delivered on CD-ROMs?" Key concepts include "multimedia," "CD-ROM," and "research." Generate a list of terms from these concepts such as

- Multimedia—slide sets, video, text, line art
- CD-ROM—CDs, online
- Research—evaluation, assessment

You could then search a database under each term and also look within each database for articles related to the other two terms. Boolean logic can help you perform the task quickly and efficiently. For instance, first you identify that you are interested in all literature on multimedia by entering your first term as "multimedia." Next, you indicate that you are interested in identifying all literature that includes both multimedia and CD-ROM by pairing the terms: "multimedia and CD-ROM." Finally, you express that you are interested in "research" with respect to "multimedia and CD-ROM," and specify a search for "research" paired with "CD-ROM and multimedia."

Searching Available Databases in Your Library

You identify the library's online catalog, online databases, CD-ROMs, and commercial databases by querying the library's literature and reference librarians. Start your search with the library's online catalog, then move to the available online databases and CD-ROM databases, and finally, if your budget permits it, consider commercial online databases. Do not consider using commercial sources until you develop skills at searching the free systems.

Using the databases selected, conduct the search using the word descriptors. Familiarize yourself with the search strategies for each available database. Some computerized systems allow you to switch between different databases as you search, which can be useful. Subsequent sections of this chapter explain online search strategies in more detail and provide guidance.

Assessing Citations

You will frequently identify a wide range of literature citations under your terms, oftentimes in an entirely different field than your topic. You must then read the entire citation—titles, abstracts, or summaries—to assess whether or not a literature citation may be useful. As you read, ask such questions as

- Is the publication/article relevant?
- Where was it published?
- When was it published?
- Will the citation possibly help find other publications?

Frequently searches will identify articles not of value to you. For example, suppose you were interested in pollution prevention, and specifically in articles based on surveys of people's attitudes, knowledge, behaviors, and actions. Your initial research question might be, "What are the knowledge levels, attitudes, and behaviors of small and medium-sized businesses concerning pollution prevention?"

The list of search terms or descriptors might include pollution, prevention, businesses, small businesses, medium-sized businesses, attitudes, behaviors, and surveys. If you did not find articles covering surveys, you could expand the list by using synonyms such as environmental, environment, toxic wastes, and environmental hazards.

Running such a search on a bibliographic database in September 1993 identified articles that reflected environmental surveys to monitor environmental factors—such as household conditions, macrobenthic surveys, liquid-chromatography for nicotine in chemical analyses, surveys of biotic similarity and diversity—but few articles focusing specifically on pollution prevention in business. Be prepared to eliminate articles reporting research topics other than those of use to you.

When possible, begin your literature review with the most recently published materials on the topic. Such articles could provide useful literature reviews and references to early works. Be aware that for some topics, recent literature may be several years old, while for other topics, research of six months ago isn't recent enough. Research and advances in some fields move much faster than in others. For example, in the early 1990s, superconductivity research more than 12 months old was considered old.

Carefully consider where the article or book was published and by whom. Articles published in research journals are usually of more value to you than articles published in general circulation magazines and newspapers. Whenever possible, select research and academic journals over trade and special interest magazines or general circulation publications. If trade or special interest magazines give the name or location of original research, use it to trace the research literature.

When reviewing a citation, try to determine if it will help you identify other relevant literature on your topic. Most research and academic journals contain extensive literature reviews, and occasionally, bibliographies. You can use such citations to search backwards for relevant research, to identify key researchers, to identify leading journals and periodicals covering the topic, and to identify related terminology for searches.

While some trade and special interest magazines will contain reference sections, most do not. You can, at times, find references to researchers in specific articles. For example, *National Wildlife* magazine carried "Mass Appeal" (Barash 1993), which reported on horseshoe crabs coming ashore to breed and lay their eggs on shores of Delaware Bay. In the article, Barash identified several scientists who have studied horseshoe crabs, including Bob Harrington, Carl Shuster, Mark Botton, and Bob Loveland. While articles such as "Mass Appeal" may not be an appropriate citation for your research, learning the names of the researchers can help you search for appropriate academic and research articles.

Once you have found the article in the library, you'll need to quickly assess the quality of the content, and determine whether the article is worth adding to your reference list. Chapter 6 provides guidance on making those educated judgments.

Printing or Downloading Citations

If you judge the citation useful, either download it to a floppy disk or print a hardcopy. Keep in mind that downloading to a floppy provides an ASCII file that you can later recall and use in preparing literature citations and literature reviews. If you anticipate downloading files, check to determine the disk size and density required for the library terminals. High density disks will hold more information and citations than lower density disks.

Retrieving Sources

Once you have acquired your citations, you must locate the materials, retrieve, and review them. Chapter 6 provides detailed guidance on retrieving references and evaluating them as well as guidance on managing those references. Keep in mind that extracting subject-specific references from literature is a recursive process in which you may find yourself returning to databases to conduct searches on related terms.

ONLINE CATALOGS AND DATABASES

Most libraries have computerized their card catalogs, which are accessed through computer terminals. The library's computer terminals may be located in one central room or scattered throughout the library. Online catalogs may also be accessible from personal computers in your office or home. While most systems are easy to learn, libraries usually provide instructions and workshops designed to familiarize users with each system's capabilities, its strengths, and its weaknesses. When you are new to a library's online system, ask for the instruction sheet explaining how to use the system.

Some systems provide access to the online catalog of other libraries as well as online databases. Such systems provide easy access to searching millions of records or files identifying different publications, articles, and other sources. Which systems are available to you will depend upon your library. Consider three online systems: OCLC, RLIN, and CARL.

OCLC

The Online Computer Library Center began as the Ohio College Library Center. In 1967 it served 54 Ohio college and university libraries within a computerized library system, and by 1993 it served more than 15,000 libraries, educational organizations, and users worldwide. Libraries that catalog their holdings on the OCLC system gain access to more than 27 million items from a range of libraries including academic, research, medical, federal, public, state, law, theological, school, community/junior college, and corporations. Some 5,500 libraries also cooperate in an electronic network of interlibrary loans. The OCLC Reference Services Database provides access to over 60 commercial online databases covering a diverse range of subjects in such areas as science, technology, business, social sciences, and humanities. In addition, OCLC provides a wide range of services to libraries to support their online catalogs and databases.

OCLC also provides FirstSearch, an online access program that provides easy access to more than 30 million records and more than 40 specific electronic databases. Users need no training; they simply follow the online instructions. In addition, FirstSearch provides a printed manual, *QuickSearch Guide*, that walks users through the basic steps of using its world catalog and databases.

RLG and RLIN

The Research Libraries Group, Inc. (RLG), consists of universities and other institutions working together to improve the access to research and learning information. Among RLG's services are Ariel, which provides document transmission service for the Internet, an information highway; CitaDel, which provides eleven citation files and document delivery service; and Eureka, which provides a simplified interface to access the Research Library Information Network (RLIN), RLG's primary service.

RLIN is an international database of bibliographic information on more than 55 million items from hundreds of research libraries, archival repositories, and academic, public, and special libraries. Its files and their extensive indexes provide speed and flexibility in research.

RLIN includes works cataloged by the Library of Congress, the National Library of Medicine, and the Government Printing Office, updated weekly; British National Bibliography records since 1984, updated regularly; comprehensive representation of books cataloged since 1968 and rapidly expanding coverage for older materials; information about nonbook materials ranging from musical scores, films, videos, serials, maps, and recordings, to archival collections and machine-readable data files; online access to special resources, such as the United Nations' DOCFILE and CATFILE records and the Rigler and Deutsch Index to pre-1950 commercial sound recordings; special files offering access to scholarly information in specific subject areas; the Library of Congress' Name Authority and Subject Authority files; and the Art and Architecture Thesaurus.

CARL

The Colorado Association of Research Libraries (CARL) provides general online access to research libraries at major universities as well as access to the online catalogs at such university library systems as Arizona, Maryland, California, and Hawaii. As of early 1994, CARL provides users access to more than 6,250,000 records of books, journals, periodicals, and other materials, as well as access to UNCOVER, an online database covering 14,000+ academic and research journals; the Magazine Index & ASAP covering 400 specialized magazines; and the Business Index and ASAP covering 800 trade, business, and related periodicals and business articles from some 3,000 additional periodicals. By late 1993, these indexes contained more than 4 million articles.

CARL also offers access to the following groups of databases:

1. Library catalogs of 22 Colorado, Wyoming, and area libraries, which include the Colorado School of Mines, University of Colorado at Boulder, University of Colorado Health Sciences Center, University of Colorado Law Library, Denver Public Library, Denver University, University of Wyoming, Colorado State University, Regis University, and the University of Northern Colorado.
2. Current article indexes and access, which includes such databases as UNCOVER, ERIC, British Library Document, Magazine Index & ASAP, Business Index & ASAP, Expanded Academic Index, and Online Libraries.
3. Information databases, which include such sources as Choice Book Reviews, Encyclopedia, Metro Denver Facts, School Model Pro-

grams, Internet Resource Guide, Department of Energy, Journal Graphics (television/radio transcripts), Company ProFile, Federal Domestic Assistance Catalog, and Librarian's Yellow Pages.

4. Other library systems, which include Boulder Public Library, MARMOT Library System (CO), Denver Public Schools (CO), Pikes Peak Library System (CO), University of Hawaii System, Montgomery County Dept. Public Libraries (MD), Houston Area Library Automation Network (TX), Inland Northwest Library, Northeastern University (MA), Sno-Isle Regional Library (WA), Arizona State University, Northern Arizona University, MELVYL (Univ. of California System), and the University of Maryland System.
5. Library and system news, which includes general information about how to use CARL and its databases, informational updates and related information, library information, and related background information on CARL activities and changes.

Using Online Catalogs

The key to effective use of any of these systems is learning the system quickly and then learning efficient search strategies. To learn the systems, ask whether your library provides orientation workshop/seminars, classroom instruction, audiovisual aids introducing services, online aids, or printed handouts. If your library offers a workshop on electronic searches, invest time in it. You'll quickly develop a basic understanding of how to operate the system, get to know its strengths and limitations, and learn useful tips for using the system.

Some library staff members will give computer orientation and instructions on electronic searching to specific users. For example, University of Wisconsin librarians provide such services to professors from across the disciplines on the UW-Madison campus.

Always check to see if the library provides handouts, pamphlets, or instructions to guide you through your initial electronic searching. From the instructions, determine which programs you can use and if the systems are fee-based. Even if the systems are free, you may need a computer account number, identification number, or password to access the computer system. Some libraries provide users free access to all services, but limit access to selected databases to reduce costs.

When learning and using the system, you need to answer the following:

- How do I log on and search the online catalog?
- How do I log on and search the databases?

- What are the limitations of the system?
- How do I evaluate an identified item?

Logging On and Searching an Online Catalog System. When you approach a terminal for the first time, check for nearby instructions, usually a card of basic commands to get you started. Follow them and log onto the system. Usually a series of screens will take you through the basic commands. In addition, the help function is available on most online catalogs as a function key, such as F1.

If the library has not provided instructions, you may find them online—i.e., on the computer screen. If the screen is blank, simply touch the space bar or any key to activate the screen. Some users may leave the computer in the middle of a search. If you start to use a machine and it has been left in the middle of a search, look for posted or online instructions. The CARL system, for example, uses "q" for quit, "x" for exit, and "s" to switch databases, and the commands are displayed along the bottom of the screen. Try the commands to see if you can return to the beginning screen. You can't harm the system by trying. If you can't figure it out, or you encounter problems, ask a librarian for help. Libraries usually have a staff member nearby to help users.

Some online systems may include several library card catalogs and databases. For example, CARL lists five general categories: 1) Library catalogs, 2) Other article indexes and access, 3) Information databases, 4) Other library systems, and 5) Library and systems news.

To select a catalog, you would enter the number "1" and then type a carriage return. The system takes you to a second screen where you select a particular catalog by entering its number, followed by a carriage return. CARL then provides an opening search screen. To proceed, you enter "W" for a word search, "N" for a name search, "B" for browsing, or "S" to switch databases. If you select "Word" search, the computer prompts you to enter the word for searching the system. Enter one of the descriptive or key terms from your research question(s), followed by a carriage return. CARL then searches the database for the books, monographs, or other holdings the library has on that term and reports the number of volumes on your subject and variations on the subject. For example, searching for "multimedia" identified some 26 books and publications in Colorado State's Morgan Library.

At this point, you can narrow the search by adding additional descriptors and creating another search. The list of items found will be displayed in a numerical list. CARL shows only seven items on a screen. You must page down for additional items.

To learn more about one of the items, you would enter its number and a carriage return, which will provide a full citation and details on the item, including its author, title, date of publication, publisher, call number, location in the library, list of descriptor terms, and whether the item is checked out or not. Acquire this information by printing, downloading to a floppy, or making a note.

When you identify a potentially useful literature citation, look carefully at the list of descriptor terms, other entries, or related terms. That list reflects the terms under which the librarians cataloged that particular volume. By reentering the term(s) in the online database, you can expand your search and are less likely to miss key information.

If you know the name of the author, or you need to search for the name of a specific individual, you can use the Name search function. This search process also provides an overall list and gives you the option of narrowing the search to more specific terms.

The Browse function allows you to search by title, call number, or series. If you enter a title, it must be the exact title. The Browse function is also handy when you are trying to clean up your citations. For example, assume that while you are doing your final citations, you realize you lack complete citation information on a particular book, but you do have the call number. You can enter the call number in the Browse function to retrieve the complete citation.

Keep in mind that the foregoing discussion provides an example of just one online system. Although other systems may operate in a similar way conceptually, the specific keystrokes vary. Learning the basics of many systems takes only a few minutes, but becoming a proficient online user takes practice. The University of Wisconsin—Madison provides a handout detailing basic search strategies (panel 5.1). Whatever the system that you use, look for such handouts and online guidance to help you.

Using Online Databases

Beyond the card catalog, many library systems provide access to a variety of online databases. For example, the CARL system provides users with such databases as UNCOVER, ERIC, Magazine Index & ASAP, Business Index & ASAP, The Expanded Academic Index, and British Library Document Supply Centre. To illustrate the range of content covered, consider each database in detail.

By late 1993, UNCOVER contained more than 4 million articles abstracted from about 14,000 academic journals with 3,000 and 4,000 abstracts being added daily. An initial search on UNCOVER is an effective starting point for researching many research and academic topics. UNCOVER uses a search strategy similar to CARL's, i.e., word, name, or browse search.

ERIC, an education database, includes the Current Index to Journals in Education, which covers professional education journals, and Resources in Education, which includes unpublished educational documents and reports available on microfiche. In addition to the search functions of word, name, and browse, ERIC provides additional search functions of subject word, subject browse, and name browse for alternative searching strategies. Online help and menus provide additional specifics. Many libraries have a collection of the ERIC microfiche documents arranged by ED numbers that serve as call numbers. If your library does not have the documents you need, you can purchase them from ERIC as either hardcopy or microfiche.

The Magazine Index & ASAP indexes over 400 general interest and popular magazines reporting on arts, current affairs, consumer affairs and production, education, leisure, people, science, and travel. The Magazine Index provides full citations to the articles as well as online and fax delivery services for a fee.

The Business Index & ASAP indexes over 800 business and management trade journals (magazines) and newspapers; business-related articles from 3,000 other publications; and 150 management and computer journals. Of these, you can view articles from some 400 journals online or have them delivered by fax for a fee.

The Expanded Academic Index indexes over 1,400 scholarly journals and the *New York Times*. Updated monthly, the Expanded Academic Index covers the humanities, social sciences, political sciences, and general sciences.

British Library Document Supply Centre is an online document delivery service that currently receives about 54,000 journals. It also supplies 170,000 journals that are no longer published. You can order specific articles from the British Library Document Supply Centre for a basic fee and have the articles faxed to you.

In addition to searching for articles, several services allow you to read the articles online or order hard copies faxed to you. You must provide your credit card number, and if you want a hardcopy, you must also provide a fax number for receiving the article. Some systems also provide gateways—links to other computer systems and databases.

Keep in mind that some institution libraries restrict access to their databases to only those people affiliated with those institutions—i.e., you must have an account or password on their system. Users from other schools or institutions are denied access.

Learning a System's Limitations

Whenever you use an online catalog or database, do learn its limitations. As you become familiar with different systems, you will learn that different schools, organizations, and companies developed their systems in different ways. Questions that you should ask about any online catalog or database include

- What dates are included?

- What publications does it include?
- How are searches executed?

In developing online catalogs and databases, the producer—whether a library or commercial company—usually identifies the earliest date for which materials are included. For example, when Colorado State University first introduced its online catalogs, signs cautioned users that it only included materials starting in 1977 and directed users to the card catalog for publications of an earlier date.

When you log onto bibliographic databases available through a library's online access system, the opening screens will indicate when materials were first entered in the database, how frequently it is updated, and related information. So do spend time reviewing the opening screens and handouts describing the databases so that you understand what they include. As you search databases and begin identifying articles, you can scroll through the full search list until you find the first entry on your topic in the database.

Also learn what fields the catalogs and databases cover. With online catalogs, all holdings may be incorporated into one catalog or the same holding may be in separate catalogs. For example, some libraries incorporate their government publications; other libraries have the government publications listed in a separate online catalog. Online bibliographic databases usually identify the general areas covered, and as you search bibliographic databases you soon learn what periodicals they include. Also, as you search for literature in your respective field, you will develop a working knowledge of both the leading periodicals in your field and relevant periodicals in related fields.

Information entered into online catalogs and bibliographic databases varies. For example, online catalog entries for a single book usually include the author, date of publication, title, publisher, city of publication, and descriptors, or other terms under which the entry is indexed. Bibliographic databases include key information such as the author(s), publication date, title, periodical, volume, number, and page numbers and perhaps descriptors. In addition, some databases may provide an abstract (article summary). Informative abstracts provide details on the article while descriptive abstracts describe the article in general, almost generic, terms. In such cases, a careful reading may either provide all the information you need or help you decide if you want to review the full article. Keep in mind that reviewing the full article may provide additional insights beyond the abstracts and that on rare occasions the abstracts may be in error.

When searching an online catalog, look for information, help, and other descriptions explaining how the systems works. For example, the CARL system explains how UNCOVER executes its searches:

Articles retrieved in UNCOVER are now sorted according to both type of material and publication date. When a word search is executed, the system first displays those items that are titles of journals, then those that are titles of articles. The articles are now sorted in reverse chronological order. Within each publication year, quarterlies or items with a publication date other than mm/dd/yy are displayed first, followed by those with mm/dd/yy dates, and finally by those with only a year as the publication date.

CD-ROMS AND THEIR USE

CD-ROMs (Compact Disc-Read Only Memory), the 5-inch optical discs for personal computers, provide a low-cost method of publishing bibliographies, statistical databases, books, manuals, instructions, tutorials, multimedia, and other information. Why? One CD-ROM holds more than 600,000 text pages, or 650 million bytes. Thus, a 24-volume encyclopedia or a national telephone directory can be stored on one CD-ROM. Software companies also distribute software on CD-ROMs, such as Lotus's Smarthelp for Windows and Corel Draw (Parker and Starett 1992).

While most libraries provide online access to large databases with wide appeal, a few libraries are turning to CD-ROM databases to cover areas of more narrowly defined subjects.

CD-ROM Bibliographies

For library use, the first applications of CD-ROMs were electronic bibliographies—counterparts to the printed abstracts and indexes. Libraries usually provide these bibliographies to their patrons at no charge. Many libraries are canceling their subscriptions to printed abstracts as the CD-ROM versions become available. For example, the University of Wisconsin-Madison uses a system of CD-ROMs, magnetic tapes, and floppy disks that enables users to search 30 million citations.

With the number of databases and titles exceeding 5,000 (Marcaccio 1993), check your library to see what bibliographic databases are available. Libraries often publish a list of their holdings of CDs.

CD-ROM Books, Manuals, and Publications

Publishing is changing rapidly and more and more books, journals, magazines, and newspapers are being published on CD-ROMs and other electronic and optical media. The communication industry, major companies, commercial publishers, and production companies are also moving to publish books, manuals, reports, and multimedia in electronic form. In scientific and technical fields especially, more and more manuals, books, and databases will be published on CD-ROMs. Doing so saves space, eliminates the need for hardcopy, greatly reduces handling costs, and speeds the publishing process.

In addition to publishing on CD-ROM, some publishers provide the opportunity to download books from their systems if you have a sufficiently large hard disk space. In 1993, Ziff-Davis, one of the leading publishers of computer magazines, books, and products, published *PC Magazine DOS 6 Techniques and Utilities* (Prosise 1993). The 1000+ printed page book also included a floppy disk. Computers owners with adequate hard disk space and a modem could log onto the ZiffNet and download the book, which required 1.1 megabytes of hard disk space. Using a 1,200 baud rate (the speed of transferring data from one computer to another: 1,200 baud=120 characters per second), a modem would require 2.5 hours. The cost was $12.95 plus the costs of the connection time required to download the book (Reid and Hume 1993).

Clearly, Ziff-Davis has made the first step of what could be a major way of delivery of books, manuals, and multimedia in the coming years, especially if technology continues to advance and transmission rates jump. In the early 1990s, most computer modems had 2,400 baud; most computer companies today regularly advertise 9,600 baud modems.

Government agencies and companies are also publishing technical reports, databases, and other materials on CD-ROM. For example, the Government Documents Department at Morgan Library, Colorado State University, provided patrons with a list of more than 125 CD-ROMs in October 1993. The available CD-ROMs included full-text databases, bibliographic databases, census data, maps, and GIS databases. The available CD-ROMs included such titles as Aeronautical Charting Data Sampler II, full-text; Agent Orange, full-text; Agriculture and Life Sciences, full-text; Aquaculture II, full-text; Brightness Temperature Grids for the Polar Regions, full-text; Delorme Global Explore, map; Department of Defense Hazardous Materials Information System, bibliographic database; George Washington Carver's papers notes and letters, full-text; Hydro-Climatic Data Network—Streamflow Data Set 1874-1988, full-text; National Climate Information Disc, full-text; OSHA CD-ROM (Standards, Directives, Documents), full-text; USGS Quality of Water: Surface and Groundwater, full-text; and Women, Water, and Sanitation: Impacts on Health, Agriculture, and Environment, full-text.

Multimedia

While publishing books on CD-ROM saves space and reduces printing costs, CD-ROMs provide even more advantages by incorporating four-color visuals, videotape clips, and sound.

Multimedia appears to be destined to become a major media by the year 2000. Schools, libraries, agencies, and companies appear to be turning to packaging a variety of information, training, and related materials in multimedia format. Developing multimedia packages holds the potential for enhancing users' involvement in the information and thus easing learning of scientific and technical information.

Finding and Selecting CD-ROMs

To use CD-ROMs in the library you must locate the CD-ROM area or room, identify the appropriate CD-ROMs for your literature review, and then access the CD-ROM.

A few libraries may have only a handful of CD-ROMs. Others may have hundreds, like the University of Wisconsin-Madison, which provides users with a four-page leaflet identifying the 100+ titles available on campus. Most libraries do provide handouts identifying the CD-ROM abstracts, indexes, and databases.

Since most libraries base their CD-ROM collection on the needs of the major research areas or focuses, you might investigate the bibliographic CD-ROMs most relevant to your field and ask your library to begin subscribing to them. Check the major printed abstracting journals and indexes covering your field and then write or call the publishers asking whether or not they provide those publications in a CD-ROM format. Also check for CD-ROM directories and available titles and publishers. The *Gale Directory of Databases* (Marcaccio 1993) and *CD-ROMs in Print* (Desmarais 1993) both provide lists of CD-ROM titles and resources.

If you will be conducting regular literature searches in your field, you might consider purchasing a CD-ROM reader for your personal computer and buying the CD-ROMs covering the literature in your respective field, especially as the prices continue to drop following advances in the technology and mass production of some CD-ROM titles. However, abstracting journals can cost from $50 to $1,500 annually with quarterly updates. You will need to consider whether or not your information retrieval needs justify such an annual investment. Certainly the cost may be fully justified if you need easy access to the CD-ROM titles or if you and your colleagues need to search the current CD-ROM titles regularly. Carefully investigate the costs of a CD-ROM title and its value to you.

Using CD-ROM Abstracts and Indexes

Once you've identified the potential CD-ROMs, investigate further. Go to the library's CD-ROM reading area or reading room, often located near the reference desk or in the reference area. Check for computers available for accessing CD-ROMs; directions, audiovisuals or online tutorials for using CD-ROMs; and printed manuals, instructions, and thesauri for the available CD-ROMs. You may also find a sign-up system for using the CD-ROMs and readers. You should either go by the library or call in and ask to use a specific title for a 30- or 60-minute time period. Some libraries check out single CD-ROMs that you then use with the library's CD-ROM readers. Check the immediate area for instructions on loading the CD-ROM. If you need help, ask the reference librarian to show you how to load the CD-ROM.

Some libraries have the CD-ROMs available on a network so that you merely select a number representing the CD-ROM and then you access it as you would a database. While the specifics vary among libraries, searching the CD-ROM follows similar procedures (panel 5.2) and libraries also provide handouts for searching specific CD-ROMs (panel 5.3).

To better prepare yourself for using CD-ROMs, study such handouts, attend any available workshops, and work through any tutorials. In some cases, the CD-ROMs will have online tutorials that walk you through a search and familiarize you with the software you'll need to use the CD-ROMs.

As you familiarize yourself with the system, notice if the CD-ROM title has a printed or online thesaurus and the details of the search functions. Before you begin your search, review the thesaurus for synonym concepts, or terms, to enhance your search. When reviewing tutorials, manuals, or other materials, focus on search strategies, Boolean operators, terms used in fields, searching complexities, and related details.

COMMERCIAL ONLINE DATABASES

You may find commercial databases helpful in your literature searches if the online catalogs, databases, and CD-ROMs available through libraries do not adequately cover the research and trade literature in your field.

Commercial databases may help you access the specialized information you need. *Gale Directory of Databases* (Marcaccio 1993) reported 8,400 databases available by 1993, and the number increases annually, as do the number of data producers, gateways, and online support services.

The commercial databases provide such resources as reference databases that include both the bibliographic and referral databases (abstracts and summaries of nonpublished information) and source databases that include numeric data (survey and statistical data), textual-numeric data (dictionaries, handbooks, and other data), full text (newspaper articles, specifications, court decisions, and so forth), software (computer programs for downloading), and images (chemical structures, maps, photographs, logs, and other visuals) (Barg and Caudra 1991).

PANEL 5.2

Familiarizing Yourself with CD-ROM Operations

As you familiarize yourself with the different searching mechanics, keep in mind that each system operates differently. The following questions were derived from comparison of CD-ROM database commands (Ali 1990).

What search or query functions does the CD-ROM access software allow? Some systems allow free-text searching for terms and phrase searches anywhere in the record; others restrict the searching to the title, terms cited, source, authors, journals, and language.

What Boolean operators does CD-ROM access software allow? Some software allows full use of AND, OR, NOT, WITH, NEAR, NONE, IN and other operators; other software restricts the operators that you can use, and some systems may not allow Boolean searches.

What field names does the CD-ROM system use? Learn the terms or phrases that identify the fields (areas in which you locate key information) in the record such as **Ti** for title, **Au** for author, **De** for descriptors, **Ab** for abstract, and **Jn** for journal. By reviewing citations, you can usually identify terms and phrases used in searches.

Does the CD-ROM system allow complex searching? In some systems you can specify which fields you want the computer to search for your terms and concepts, and you can direct the computer to do a variety of searches in one step.

Does the CD-ROM system allow you to browse? If so, what are the limitations? Some software may restrict the way in which you browse and the nature of your browsing. For example, SilverPlatter software allows you to browse when searching in the Index functions; ISI software allows you to browse the dictionary of titles, authors, journals, addresses, and more. Other systems allow you to browse by year, subheadings, terms, and names.

Does the CD-ROM software allow truncation? Truncation is adding a * or ? at the end or beginning of a partial spelling of a term so that the system searches for a wider range of terms. Assume the system uses the * for truncation, and you're searching for the terms computer, computing, and computerization. You can enter a *comput** to truncate the term and the computer will then search for records for all terms that begin with *comput* but could end in other word patterns. While truncation may speed your search, it can create false "hits." For example, the *comput** truncation given above would also produce hits on computation, computational, or other terms that begin with comput.

Does the CD-ROM system allow combining numbers for subsequent searching? Once you complete a search, the system comes back with a count of the number of hits (identified articles or citations) as a line entry. Some systems allow you to combine the searches by reentering the lines identifying the number of hits for each separate hit. Let's say you are searching PsycLIT for the terms online and reading. You enter each one individually as a separate search:

Search Number	Number of records	Term
1.	133	Online
2.	6357	Reading

You can then combine the search to see which hits include all terms by writing a command statement as "1 and 2." The narrower search would then produce citations containing all terms.

PANEL 5.2 (continued)

FAMILIARIZING YOURSELF WITH CD-ROM OPERATIONS

How can you display the identified records? Once the CD-ROM software identifies the number of hits, or citations, you must tell the system how to display them. Generally, you will need to follow online commands to display the records.

How can you print your search? When conducting searches you may find only a dozen articles that you want to retrieve out of some 200 identified in the search. You do not need to print all 200, so you must tell the computer which one you do want to print by marking these items. Marking and printing citations is easy on some systems. You cursor down, hit the space bar to highlight the item, and then hit the print key. Again, online instructions should be available.

How do you save or download your search? On most systems, you can save the search to a formatted disk that you bring along with you. Check with the library to see what size and density disks their personal computers use. Some computers provide for both the 5.25-inch and 3.5-inch high density disks, but some may take only one size. In general, high density disks are best because they can hold more information.

Keep mind too the copyright restrictions on downloading searches. Generally the CD-ROM publisher allows downloading and printing for your educational efforts, but you cannot generate a bibliography from the database and plan to sell it.

Does the CD-ROM title require searching multiple disks? Because of the massive number of citations in fields, CD-ROM publishers often provide a series of disks, often by year, that cover the respective topic. So you may have to remove one CD-ROM and place a second in the computer, or switch, when using multiple CD-ROMs. Some systems allow you to save the search strategy to rerun on other disks. If you will be searching a title with multiple disks, begin with the most recent disk and work backwards.

What are the general commands for moving through the system? When you familiarize yourself with the commands, look carefully for the commands that enable you to return the to main menu, to access the online help, and to exit to DOS. Usually the main menu includes such commands, but the commands do vary across systems. For example, F1 invokes the help function on many, but not all, systems.

Each database provides specific guidance on the rights to its usage. Some services restrict downloading and the importing of information into subsequent databases; other services provide the rights to do so, but limit your potential use. The usage guidelines and manuals detail the usage rights of the respective information.

Libraries often subscribe to information services that provide access to some commercial databases. Such services may charge an annual fee as well as fees for hourly hook-up, printing, and downloading. These charges may be passed on to users. Such services constantly update their services, add additional databases, and provide new files for their clients.

While using such databases is becoming easier, searching of some databases is still best accomplished with the aid of a trained librarian. Vendors may also provide simplified tutorials so that researchers can conduct their

PANEL 5.3

PsycLIT SilverPlatter Quick Reference Card

PsycLIT®

SilverPlatter®

Quick Reference Guide

To restart system	Press **F7** (**Restart**)	**A phrase**	**Find:** well being
To select a database	Use arrow keys to highlight database Press spacebar, then **Enter**	**A word root**	**Find:** famil*
		Internal or limited truncation (one or no characters)	**Find:** behavio?r ; **Find:** norm?
List of commands	Press **F10**, then highlighted letter	**To combine concepts:**	
Database information	Press **F10**, then **G** for **Guide**	**Use AND to narrow search**	**Find:** symbolism and language
To use Thesaurus	Press **F10**, then **T** for **Thesaurus**	**Use OR to broaden search**	**Find:** wellness or health
To link chapters, books	Place cursor on <<SEE BOOK>> or <<SEE CHAPTER>> Press **L** (for **Link**)	**Use NOT to narrow search**	**Find:** advertising not television
		Use WITH to restrict search to same field	**Find:** crowd* with violence
To search authors	Press **F10**, then **I** for **Index** Type last name, then **Enter**	**Use NEAR to narrow search to number of words in proximity**	**Find:** computer near anxiety **Find:** expert near2 system?
To look for: **A word**	**Find:** hesitation		

FIELD NAME	SAMPLE JOURNAL RECORD	SEARCH EXAMPLES
Title	TI: Effects of rock and roll music on mathematical, verbal, and reading comprehension performance.	rock with music in ti
Author	AU: Tucker,-Alexander; Bushman,-Brad-J.	tucker-alexander in au
Author Affiliation	IN: Iowa State U, US	iowa state in in
Journal Name	JN: Perceptual-and-Motor-Skills; 1991 Jun Vol 72(3, Pt 1) 942	perceptual-and-motor-skills in jn
ISSN	IS: 00315125	00315125 in is
Language	LA: English	english in la
Publication Year	PY: 1991	1991 in py ; py=1989-1992
Abstract	AB: 151 undergraduates completed mathematics, verbal, and reading comprehension problems while listening to rock and roll music played at 80 db or in silence. The music decreased performance on math and verbal tests but not on reading comprehension. (PsycLIT Database Copyright 1992 American Psychological Assn, all rights reserved)	rock near1 roll in ab
Key Phrase	KP: rock & roll music; mathematics & verbal & reading comprehension performance; college students	mathematics with performance in kp
Descriptors	DE: ROCK-MUSIC; READING-COMPREHENSION; MATHEMATICAL-ABILITY; VERBAL-ABILITY; ADULTHOOD-	rock-music in de ; adulthood- in de
Classification Codes	CC: 2340; 23	2340 in cc ; 23 in cc
Population	PO: Human	human in po
Age Group	AG: Adult	adult in ag
Update	UD: 9201	9201 in ud
Accession Number	AN: 79-00362	79-00362 in an
Journal Code	JC: 1576	1576 in jc

FOR SEARCH HELP, CONTACT:

PsycINFO User Services
American Psychological Association
750 First Street, NE
Washington, DC 20002-4242

Telephone:
(800) 374-2722 (in North America)
(202) 336-5650
FAX: (202) 336-5633; TDD: (202) 336-6123
Internet: psycinfo@apa.org

American Psychological Association

11/93

(over)

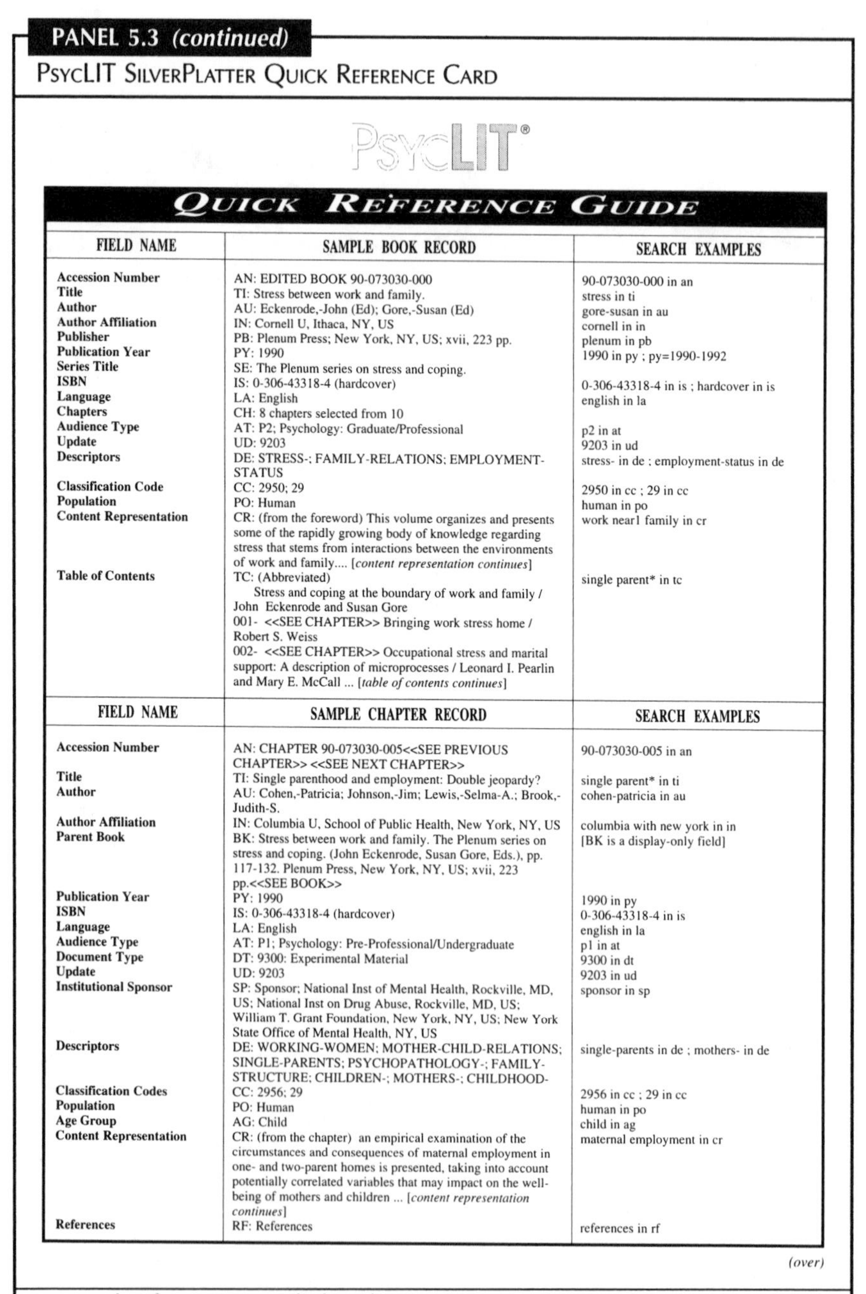

PANEL 5.3 *(continued)*

PsycLIT SilverPlatter Quick Reference Card

PsycLIT®

Quick Reference Guide

FIELD NAME	SAMPLE BOOK RECORD	SEARCH EXAMPLES
Accession Number	AN: EDITED BOOK 90-073030-000	90-073030-000 in an
Title	TI: Stress between work and family.	stress in ti
Author	AU: Eckenrode,-John (Ed); Gore,-Susan (Ed)	gore-susan in au
Author Affiliation	IN: Cornell U, Ithaca, NY, US	cornell in in
Publisher	PB: Plenum Press; New York, NY, US; xvii, 223 pp.	plenum in pb
Publication Year	PY: 1990	1990 in py ; py=1990-1992
Series Title	SE: The Plenum series on stress and coping.	
ISBN	IS: 0-306-43318-4 (hardcover)	0-306-43318-4 in is ; hardcover in is
Language	LA: English	english in la
Chapters	CH: 8 chapters selected from 10	
Audience Type	AT: P2; Psychology: Graduate/Professional	p2 in at
Update	UD: 9203	9203 in ud
Descriptors	DE: STRESS-; FAMILY-RELATIONS; EMPLOYMENT-STATUS	stress- in de ; employment-status in de
Classification Code	CC: 2950; 29	2950 in cc ; 29 in cc
Population	PO: Human	human in po
Content Representation	CR: (from the foreword) This volume organizes and presents some of the rapidly growing body of knowledge regarding stress that stems from interactions between the environments of work and family.... *[content representation continues]*	work near1 family in cr
Table of Contents	TC: (Abbreviated) Stress and coping at the boundary of work and family / John Eckenrode and Susan Gore 001- <<SEE CHAPTER>> Bringing work stress home / Robert S. Weiss 002- <<SEE CHAPTER>> Occupational stress and marital support: A description of microprocesses / Leonard I. Pearlin and Mary E. McCall ... *[table of contents continues]*	single parent* in tc

FIELD NAME	SAMPLE CHAPTER RECORD	SEARCH EXAMPLES
Accession Number	AN: CHAPTER 90-073030-005<<SEE PREVIOUS CHAPTER>> <<SEE NEXT CHAPTER>>	90-073030-005 in an
Title	TI: Single parenthood and employment: Double jeopardy?	single parent* in ti
Author	AU: Cohen,-Patricia; Johnson,-Jim; Lewis,-Selma-A.; Brook,-Judith-S.	cohen-patricia in au
Author Affiliation	IN: Columbia U, School of Public Health, New York, NY, US	columbia with new york in in
Parent Book	BK: Stress between work and family. The Plenum series on stress and coping. (John Eckenrode, Susan Gore, Eds.), pp. 117-132. Plenum Press, New York, NY, US; xvii, 223 pp.<<SEE BOOK>>	[BK is a display-only field]
Publication Year	PY: 1990	1990 in py
ISBN	IS: 0-306-43318-4 (hardcover)	0-306-43318-4 in is
Language	LA: English	english in la
Audience Type	AT: P1; Psychology: Pre-Professional/Undergraduate	p1 in at
Document Type	DT: 9300: Experimental Material	9300 in dt
Update	UD: 9203	9203 in ud
Institutional Sponsor	SP: Sponsor; National Inst of Mental Health, Rockville, MD, US; National Inst on Drug Abuse, Rockville, MD, US; William T. Grant Foundation, New York, NY, US; New York State Office of Mental Health, NY, US	sponsor in sp
Descriptors	DE: WORKING-WOMEN; MOTHER-CHILD-RELATIONS; SINGLE-PARENTS; PSYCHOPATHOLOGY-; FAMILY-STRUCTURE; CHILDREN-; MOTHERS-; CHILDHOOD-	single-parents in de ; mothers- in de
Classification Codes	CC: 2956; 29	2956 in cc ; 29 in cc
Population	PO: Human	human in po
Age Group	AG: Child	child in ag
Content Representation	CR: (from the chapter) an empirical examination of the circumstances and consequences of maternal employment in one- and two-parent homes is presented, taking into account potentially correlated variables that may impact on the well-being of mothers and children ... *[content representation continues]*	maternal employment in cr
References	RF: References	references in rf

(over)

own searches. Libraries often provide information sheets on using the commercial databases, making the needed arrangements for your search, and how the libraries charge users.

The following discussion explores the services provided by DIALOG, BRS, CompuServe, and other commercial database vendors.

DIALOG

DIALOG Information Services provides the DIALOG Information Retrieval Services with access to more than 400 databases covering diverse fields. Those of relevance to technical and scientific professionals are agriculture; food and nutrition; biosciences and technology; business; chemistry; computers and software; energy and environmental topics; engineering; government and public affairs; medicine and health; patents, trademarks, and copyrights; physical science and technology; and the social sciences and humanities.

DIALOG also provides an electronic mail service; an online document ordering service of original publications; online full-text sources; CD-ROMs of selected databases; and other services, including online viewing of articles from more than 2,500 magazines, newspapers, trade journals, newsletters, and market research reports.

DIALOG charges depend upon the services provided. Subscribers pay an annual fee as well as a use fee. For each search, users pay their local telephone company a connect fee and pay DIALOG both a connect fee and a per citation fee. Information about DIALOG service is available from DIALOG Information Services, Inc., Marketing Department, 3460 Hillview Avenue, Palo Alto, CA 94304; telephone 800-3-DIALOG or 415-858-3785.

BRS

InfoPro Technologies provides access to about 150 bibliographic and full-text databases through BRS Search Service and access to more than 100 databases through BRS After Dark. BRS Search Service is available to provide information on a regular basis throughout the day. BRS After Dark is available from 6 p.m. to 4 a.m. Monday through Friday, 6 a.m. to 2 a.m. Saturday, and 9 a.m. to 4 a.m. Sunday.

The databases include medicine and pharmacology; education; life sciences; physical and applied sciences; references and multidisciplinary; business; social sciences and the humanities; and special BRS databases.

Charges for using BRS include a telecommunications charge and an annual charge paid either in advance or on a monthly rate. BRS After Dark charges include a one-time subscription fee, a minimum monthly charge, an hourly access charge, and a display charge per citation. For information on BRS and BRS After Dark, contact BRS Online Products, A Division of InfoPro

Technologies, 8000 Westpark Drive, McLean, VA 22102; telephone 1-800-955-0906.

CompuServe

CompuServe provides access to more than 850 different reference, or bibliographic, databases through the IQUEST service. IQUEST provides access to a wide range of databases such as accounting, advertising and marketing, aerospace and defense, agriculture and food, biology and biotechnology, chemistry, civil engineering, computer science, data processing and transmission, earth sciences, electrical engineering, energy, engineering, environment, government, life sciences, materials science, medicine and health care, pharmacology, physics, psychology, science, social science, telecommunications, and transportation.

In addition, CompuServe provides a full-text search to more than 700 professional journals, newspapers, consumer magazines, specialized newsletters, and published research papers. Information provided to subscribers details the processes for accessing reference databases.

Beyond the regular monthly fee, CompuServe charges a connect time for the different databases as well as a charge per citation listed. Charges vary from database to database, from $2 to $75 per database and $1 to $2 per article/listing. Charges are billed automatically to your account. For information, contact CompuServe, P.O. Box 2012, Columbus, OH 43220; telephone 1-800-368-3343, Ext. 23.

INFOTRAC

Updated monthly, the database of InfoTrac EF General Periodicals Index Academic Library Edition covers about 1,100 general interest and scholarly publications and spans the last three to four years. Other databases available with InfoTrac EF include an Academic Index™, Expanded Academic Index™, National Newspaper Index™, Business Index™, F&S Index plus Text™, Health Index™, LegalTrac™, Magazine Index Plus™, General Business File™, and InvesText™.

Emerging Systems

New online systems and databases will emerge to further enhance the searching in respective fields.

For example, the Institute of Scientific Information distributes the Science Citation Index Compact Disc Edition with Abstracts (SCI® CDE with Abstracts) and the Science Citation Index Compact Disc Edition (SCI® CDE). Subscribers receive easy-to-use software that enables diverse searching of the CD-ROMs. The SCI® CDE with Abstracts provides full-text, English-language abstracts written by the articles' authors. Users can invoke the Keywords and KeyWord Plus for identifying additional articles related to searches.

The Internet

The Internet is an existing information highway that evolved from a Department of Defense electronic network. Today, the Internet consists of dozens of networks linked together. For many years, Internet users were primarily scientists, students, and university personnel, but now many more users can gain access through commercial systems. Internet users can send electronic mail around the world and can access hundreds of online catalogs of hundreds of universities and colleges as well as electronic databases. Appendix A provides a closer look at the Internet and its available resources.

REVIEWING THE SYSTEMS FOR ACCESSING ELECTRONIC INFORMATION

As mentioned earlier, you may be able to access electronic systems though computer terminals located in the library or, in the case of academic libraries, other computer laboratories on campus; or through office and home computers equipped with a modem and software.

Once you identify a citation of potential value, you need to record the salient information so that you can search the library and obtain it. Your approach depends upon which method you are using to access the materials.

Using the Library's Computer Terminals

If you are using online databases in the library, or through an already networked system, find out if the system enables you to print and/or download identified citations. Libraries may have computer terminals with disk drives for downloading your information, printers, or dumb terminals. The best arrangement is personal computers with standard disk drives and attached printers that are networked directly to the computer system. In such cases you can either print the citations through a series of simple commands, or you can download the citations to a floppy disk.

Printing is usually straightforward. In most cases you will find instructions either online or nearby. Before you begin, check to see that the printer is on and that it has adequate paper for printing the citations. If not, ask a librarian to show you how to use the system or to load the needed paper.

If the computers have disk drives for downloading the key information, make sure that the disks you bring are the right size and density. Most systems use either 5.25-inch or 3.5-inch high density floppy disks. High density floppies provide the distinct advantage of allowing you to minimize the number of disks you need. For example, a high density 5.25-inch floppy holds nearly the equivalent of four low-density disks; a high density 3.5-inch floppy holds the equivalent of two low-density disks.

Some libraries use IBM computers; others may use Apple's Macintosh computers. If they are older models, then you may have other compatibility problems. When you download the files, the systems ship them as ASCII files that should be readable in DOS by full function wordprocessing programs such as WordPerfect, MS Word, and WordStar in both the DOS and Windows versions. Often, you may find it necessary to reformat the pages and adjust lines, margins, and spacing.

Using Your Office or Home Computer

You can access online catalogs and databases through personal computers attached to either optical networks or the telephone lines. To do so requires different hardware configurations and equipment. The following discussion briefly reviews the needed equipment, software, and access programs, and provides suggestions for searching from your office or home personal computer.

More and more universities, colleges, businesses, and industries are networking all personal computers with optical cable. Such systems allow e-mail, easy transfer of data and files between computers, and searching online catalogs and databases.

To use a personal computer on such an optical network requires a network card, cabling to the network, and communications software. In most settings, the organization will have technical specialists who install the cable, add the card to the personal computer, install the needed software, and test the system. If you have access to a personal computer on such a system, find out how to use the software, including how to access the online catalog, how to capture files and download them to both hard and floppy disks, and how to print the literature you identify with a search.

Your personal computer will need a modem, a nearby telephone line, communication software, and surge protectors for both the electrical and telephone lines. Modems transfer files between one computer and another at varying speeds: 1,200, 2,400, 9,600 baud, and faster. Since ten baud equals about 1 character, a 1,200 baud modem transmits about 120 characters a second.

You will need to know 1) the telephone number to call to access the online catalog or the number to access the computer, which in turn accesses the online catalog, and 2) the required settings for communicating with the computer—i.e., the baud rate, parity, data, stop, and duplex. You must set the modem on your computer to match these values so that your computer and the computer you access can talk to each other. Without the same settings, the computers will not be able to understand one another or communicate key information. Most libraries have handouts giving directions for setting up your modem to access their respective online systems.

If you are new to using a modem, do not underestimate the time required to set up a personal computer and modem. Although installation could go quickly, it could also require several hours over a couple of days. For example, when upgrading to a 486 computer and a Windows communication package in early 1993, we found ourselves calling the communications software manufacturer to troubleshoot the problems. Only when we called the computer manufacturer did we obtain the needed instructions for resetting the machine code so that the communications software would work.

Once you have the basic connecting setup, you will need to enter the proper telephone number, baud, parity, full or half duplex, and other settings for your modem, and then test the system. If everything works, you dial the access computer, log in, and then follow the directions for your respective searches.

If you plan to download citations to a hard or floppy disk on your computer, review your communications software manual. When downloading to a disk, you usually need to specify the preferred disk drive and may also have to provide related information. Before doing so, make sure the disk is formatted and that you have plenty of space for the needed information.

When downloading information, the system creates an ASCII file that you can then retrieve into many standard word processing programs. When you pull the file up, you will find you will need to edit it to reduce its length and remove unneeded information. If you plan to print the file, consider using a smaller typeface to compresses the information onto fewer pages. To remove the excessive carriage returns in the file, you can use the search and replace function of your word processing software.

Developing online searching skills begins with understanding your problem, being able to articulate the information you need, finding the key reference sources, using those sources to identify the needed information, and then retrieving that needed information. All are skills that you can develop over time. This chapter has provided an introduction to the information and skills that will serve you well as you begin developing these skills.

REFERENCES

Ali, S. N. 1990. Retrieval commands of CD-ROM databases: A comparison of selected products. *CD-ROM Professional*, May 1990: 28–33.

Barash, L. 1993. Mass appeal. *National Wildlife* 31(4): 14–19.

Barg, J., and R. Caudra, eds. 1991. *Directory of online databases*. Detroit: Gale Research.

CompuServe. 1993. *CompuService product information brochure*. Columbus, OH: CompuServe.

Desmarais, N., ed. 1993. *CD-ROMs in Print*. Westport, CT: Meckler.

Marcaccio, K. Y., ed. 1993. *Gale directory of databases*. Detroit: Gale Research.

Parker, D., and B. Starett. 1992. *Technology edge: A guide to CD-ROM*. Carmal, NJ: New Riders.

Prosise, J. 1993. *PC Magazine DOS 6 Techniques & Utilities*. Emeryville, CA: Ziff-Davis.

Reid, T. R., and B. Hume. 1993. New PC books on the way, but how do you curl up with a good disk? *Rocky Mountain News*, Sunday, 4 July 1993, Business section, p. 91A, col. 1–3.

CHAPTER

6

Using the Literature

Once you identify relevant literature, a systematic plan can speed your process of obtaining scientific or technical information, minimize mistakes, and boost your chances of obtaining the subject-specific literature you need.

Efficient literature use includes

- Managing citations and reference materials
- Retrieving publications
- Assessing and extracting information
- Creating a bibliographic database
- Considering ethical and legal issues

MANAGING CITATIONS AND REFERENCE MATERIALS

If you conducted electronic searches, you either printed your search results or downloaded the literature to a file. To prepare printed literature citations for retrieving, cut your citations into individual sheets. Make sure that each citation has its call number as well as its location in the library. If you do not have that information, obtain it. Check the library information sheets for the locations of the publications within the library.

If you prepared notecards, make sure that you have all of the relevant information as illustrated in panel 6.1. Keep in mind that you will need all available information when you write your reports or articles, or document your references. If you are preparing the notecards on a computer, prepare a template with blanks for the needed information and then copy that template as many times as needed for your cards. In some word processing systems, you can prepare a macro command that contains the information.

PANEL 6.1

AN EXAMPLE OF THE FORMAT FOR A NOTE CARD

Author:

Date:

Title:

Publisher:

City:

Key terms:

Source of the citation:

Call number:

Library location:

Original source:

Add your name and telephone number to your notes. Should you lose them, someone could find them and call you. Such a precaution can preclude many hours of searching and digging. To keep your notes organized, punch a hole in the upper left-hand corner, organize them by call number order, and keep them in sequence with a split ring. Should you drop the cards, they will not become jumbled and disorganized.

Managing References

As you begin collecting your literature, develop a way of managing the files and organizing the books and reports for easy reference, such as developing a file system for photocopies, reprints, and notes, and reserving a shelf or shelves for books and reports. In the long run, an organized filing system is an effective management tool that saves you time, effort, and possibly money.

Filing systems can be very simple to extremely complex. A $15 investment in a plastic filing crate, hanging file folders, and interior folders (available in office supply or discount stores) provides an effective way of storing, transporting, and handling the references. As projects become larger, you can move the references to standard office file cabinets.

Careful labeling of file folders is critical to finding your notes and literature. Two common methods include 1) writing the last name of the author(s) and the date of publication on the file folder tabs, and 2) giving each citation a unique access number and writing that number and the author's last name and publication date on the tab. In the first method, organize the file folders in alphabetical order by the senior author's last name; in the second method, organize the file folders in ascending order of the bibliographic numbers on the file folders.

Keep books, reports, and other publications on a bookshelf, or part of a bookshelf, and keep all publications together. Organize the publications so that all library books are in one location and personal copies in another location. Organize the library collection by call numbers so that you can easily find a volume, should the library request that you return it.

RETRIEVING PUBLICATIONS

Whether you are retrieving books, periodicals, or government publications, your retrieval will benefit from a systematic approach.

Retrieving Books and Monographs

Working with your list of books, first determine whether your library has the book or not. In your initial search, you should have noted each book's location in the library, so you need only sort out the citations that you think the library does not have. Double-check with another catalog search.

After compiling the list of library books to retrieve,

1. Sort the books and monographs by call number
2. Group all citations by call letter, location, and floor
3. Go to stacks and locate the publication
4. Make a preliminary assessment of the publication
5. Check out the desired publications

When sorting the citations by call number, check to make sure that all publications with the same call numbers are housed in the stacks together. Some libraries have an oversized publication section for publications that exceed a specified size—for example, greater than nine inches tall. Check your library's guide to stack publications to determine where the publications are housed, and then organize your citations by floor or library area.

If you anticipate checking out several volumes, bring along an attaché case, briefcase, book bag, or backpack to carry the publications. Dress appropriately. Many people work up a sweat as they move from one floor to the next and hurry through the stacks.

Go to the specific floor of the library and scan the call numbers on the stacks quickly to identify the general area of the publications you are seeking. Narrow your search to the specific call number, using the sequence of letters and numbers to guide your searching.

When you remove a publication from the shelves, do not reshelve it. Leave it on a nearby table, library cart, or shelves designated for reshelving books. If you reshelve the book in the wrong place, others will not be able to find the publication.

If you can't find the publication, double-check your search strategy. Are you working in the right sequence? Are you reading the call numbers properly? Check the jumps within the shelves. Are they in sequence? Do you find any "flags" telling you to search in a different area or an alternative location? Such a flag might be a piece of wood, marked with the call numbers of the missing volume and providing instructions on where to look, such as in the reference section of the library or at the reserved reading desk.

If the book is not in the stacks, double-check the call number on the card or online catalog. Some electronic systems indicate if a book is checked out. You could also ask the checkout clerk. The library may recall a checked-out book for you. If it should be on the shelf but isn't, check the carts of books awaiting reshelving and the list of books in storage, if the library places older volumes in storage. If you fail to find the publication, then go to the checkout desk. Ask the clerk about services the library might provide to check for missing publications or for books in storage.

Once you find the publication, scan the table of contents, executive summary, chapter summaries, notes on the author's background, and other pertinent information to determine whether the publication will provide the information you need. A later section of this chapter provides guidance on assessing publications and extracting information.

Retrieving Periodicals

To retrieve periodical articles

1. Sort by periodical title
2. Arrange from most recent to oldest
3. Obtain call number
4. Check the library's current holdings with the online catalog or card catalog
5. Check the periodicals room for current issues
6. Check the periodical stacks for the back issues

After you sort the citations by periodical titles, arrange multiple articles from the same publication from most recent to oldest. To determine if your library has your citations, always check the card catalog or the online catalog.

University research libraries usually maintain a large collection of the periodicals most used by students, faculty, and researchers. Keep in mind, however, that all libraries sometimes cancel periodicals. Therefore, check a library's holdings before trudging through the stacks. If the online system does not provide a record of the dates of the library's holdings, check the serials, or periodicals, card catalog to see what years the library has. Keep in mind that some libraries place the older volumes in storage.

If your library has a periodicals reading room, check the current issues for the articles you need, and then go to the stacks. If you do not find the volume and issue you need, check any tables, desks, photocopy areas nearby, and the reshelving carts, remembering that libraries usually suggest that users not reshelve any book or periodical that they use. If you still cannot find a current volume from the last few years, check with the library staff to see if the issues are being bound and, if so, when they will be available.

For older volumes, check to see if they are in storage. If you cannot locate the back issues, check with the reference librarian and checkout clerks for help. Some libraries allow users to check out older bound volumes. In other cases, a user may have taken the bound volumes to another floor to read or to photocopy selected articles. Library staff members pick up volumes throughout the day, return them to the reshelving carts, and then a staff member reshelves the volumes. If you still cannot find the volume, you can check again in a day or two.

When you do find the volume you need, make sure it contains the article that you need. Citations, even in the best publications, can include the wrong volume number, date, page numbers, or publication title. If the volume does not contain the cited article, first check for confusion on the publication date and volume numbers by comparing the date and volume numbers of other bound issues. For example, if you are looking in volume 15 for an article with a publication date of 1990, and you find volume 15 has a 1985 publication date, you know the citation is in error. In this case, check the volume with the 1990 publication date.

Retrieving Government Publications

Retrieving government publications can be one of the more difficult tasks in a citation search. You should use a systematic approach.

1. Check your library's catalog(s)
2. Sequence the citations to the library's holdings by SuDoc call number
3. Identify the area(s) in the library with government publications
4. Locate the publications

If you encounter problems searching the stacks for government publications, look carefully at the call numbers, and double-check the call numbers for those you cannot find. If you still have problems, ask the librarian who specializes in government publications for help. Keep in mind that the same topic may be covered by different agencies, government agencies often change names, research responsibilities often rotate between agencies, and the government forms new agencies and consolidates scientific and technical activities.

Using Interlibrary Loan Service (ILL)

Interlibrary Loan enables you to obtain information that your library does not have. To use the ILL service, you must identify the needed publication or article, check your library's online database and catalog(s) to make sure your library does not own the item, and fill out a request.

ILL materials can arrive from a few days to over a month after your request, so begin your literature search early in the term and request information early in the semester. If you are a researcher, such delays seldom create problems. If you are in a rush, consider having an electronic delivery service provide you the article by fax, or read the article directly online. Before you order a fax, check the fees. Many articles cost between $5 and $15; others may exceed $100.

If you do request publications or articles through ILL, keep a record of the ILL requests. Note the date of your request, and when your interlibrary loans arrive, check them off. This will prevent duplicate or overlooked requests.

ASSESSING AND EXTRACTING INFORMATION

Once you obtain salient literature from your articles, books, and publications, your task shifts to literature evaluating and managing.

Evaluating and managing the literature you retrieve includes reading literature purposefully and taking careful and thorough notes.

Reading Literature Purposefully

Two philosophies exist for reading literature with the resolve to evaluate and extract specific references: the in-library process and the in-office or at-home process.

The in-library process includes reading literature thoroughly and extracting specific references so that you never need to review the literature again. This philosophy, more common in the humanities, works well if you know exactly what you want and will not need the literature to extract further references. The in-office or at-home process calls for quickly assessing literature in the library, then checking it out or, as permitted under copyright law, photocopying as needed. Then, when you are in your office or at home, you carefully extract salient references.

The second philosophy, more common in the sciences, works well when you are faced with investigating a less lucid problem. In such situations, you may need to build a library of photocopied articles, abstracts, article reprints, books, and reports. On first reading the literature, you may be seeking only references on the research question, methods, and outcomes. As you develop a deeper understanding of the problem, you may need to return to the literature to extract methodological details, a better understanding of the findings, or a review of the recommendations in your literature.

Your selection of either the in-library or the in-office or in-home reading process depends upon your reading motive. Huckin (1983), suggests that five different reading "styles" exist for reading scientific and technical materials. These include

- Skim reading to learn the general drift of a passage
- Scan reading to quickly find specific information
- Search reading with attention to the meaning of specific terms
- Receptive reading to fully comprehend the materials
- Critical reading to evaluate the content

Huckin offers that each reading "style" may be useful at different times in the literature evaluating process. Skimming and scanning could be appropriate strategies when determining whether or not you should check out a book or make photocopies of an article. Search, receptive, and critical reading may be appropriate when carefully reading the materials and taking notes.

These multireading strategies may also be useful when evaluating different types of literature. Consider applying these multireading strategies to

- Academic and research journal articles
- Trade and special magazine, newsletter, and newspaper articles
- Newspapers and popular periodicals
- Textbooks
- Books, monographs, and reports

Reading Academic and Research Journal Articles. In scientific and technical fields, different researchers may define terms differently, use different measurement strategies, or use different terminology. Search reading would uncover such differences. Once you understand the terminology being used, receptive reading could provide a foundation for critical reading.

When critically reading research articles, ask questions such as

- What is the author's expertise?

- What is the rationale, justification, or theoretical basis for the project?
- What is the date of the publication?
- Is a literature review included? If so, is it thorough?
- What research questions, objectives, or purposes were proposed?
- What methods were used to collect the information?
- Are the methods standard for the research in the discipline?
- How many objects, animals, plants, or people were studied?
- How were they selected?
- What data analysis methods were used?
- Are the data analysis methods standard for research in the discipline?
- What were the primary findings?
- What limitations did the authors place on the findings?
- What suggestions did the authors make for future studies?

By asking such questions, you begin to develop a critical perspective with which to evaluate the quality of the research. As you develop a clearer understanding of the research methodologies, theoretical perspectives, and overall paradigm of your field, you will begin asking more critical questions. As you read, take notes that reflect answers to your questions.

Literature review articles often consist of extensive, and sometimes exhaustive, reviews of the major research topic or more narrowly defined problem, topic, or subject area. For example, technical communicators and software developers would find several publications on hypertext, hypermedia, and cognitive science especially helpful. Nielsen (1990) provides a careful review of the thinking, research, and developments of research on hypertext and hypermedia through 1990; Osherson and Lasnik (1990), Osherson, Kosslyn and Hollberbach (1990), Osherson and Smith (1990), and Baraslou (1992) provide overviews of cognitive science.

When critically reading literature review articles, ask such questions as

- What is the author's expertise?
- What was the rationale for conducting the literature review?
- What specific topics were included? Why?
- What specific topics were excluded? Why?
- What research questions, objectives, or purposes were proposed?
- Did the authors report research from the last two years? Five years?
- What methods were used to collect the information?
- How many articles were reviewed?
- How were the articles selected? What rationale did the author offer?

- Did the authors critically review the articles' content?
- What criticisms were offered of the articles?
- How did the author organize the literature review article?
- Does the organization reflect major concepts in the discipline?

Such questions provide a foundation, or framework, for reading literature review papers. Keep in mind that only one or two sections of an exhaustive review may be useful to you. Especially note the journals and publications in which cited studies were reported. If you need to research the topic further, consider that issues of those particular journals may carry more recent articles on your subject.

Reading Trade and Special Magazine, Newsletter, and Newspaper Articles. Most trade and special publication articles provide practicing professionals with information designed to help them improve their knowledge, improve their skills, and better their problem solving abilities. These articles are written by practicing professionals, researchers, reporters, writers or editors with specialized backgrounds and experience, or by general assignment writers and editors. Thus, the quality and thoroughness of the information varies widely.

Many authors base their articles on interviews with professionals and researchers, and literature reporting recent research, but other authors may be less thorough. Begin evaluating these articles by asking such questions as

- What seems to be the overall quality of the publication?
- Does the periodical have a "respectable" reputation in the profession?
- What key points does the author make?
- What is the author's background?
- What are the sources of his/her information?
- Does the article fully identify researchers or experts quoted?
- How many information sources appear in the article?
- Does the article cite specific publications?
- Does the literature mentioned have full citations?
- Does the article present both sides of controversial subjects?
- Does the article favor one viewpoint? Slant the presentation?
- What kinds of evidence does the article use? Observations? Interviews? Reviews of research studies?

No one set of general questions can cover all topics, so read with a questioning mind. Again, as you become familiar with the topic, you can develop a more critical reading style.

Be especially alert to any mention of or references to researchers, experts, research projects, research reports, journal articles, and other information sources on the topics that you are investigating. Prepare note cards on such identifications and use the information to conduct searches of abstracts, indexes, electronic databases, and CD-ROMs on your respective topic. This will begin a networking process that can help you identify relevant information that you may have missed in your initial literature search.

Reading General Newspaper and Popular Periodical Articles. Some general newspapers and popular periodicals provide specialized sections focusing on specific topics. Articles tend to report current issues, developments, and progress on broad-ranging topics. When gathering information for these articles, reporters often turn to professionals, experts, and administrators. Some specialized writers depend heavily on news releases issued by research institutions and professional societies to alert them to developments in scientific, technical, and specialized fields. For example, many newspaper articles on health research are based on articles from such leading medical journals as *The New England Journal of Medicine*. While some reporters may be specialists in science, medicine, environment, or business, many writers and editors are often generalists.

When reading general newspaper and popular periodical articles, ask

- What are the main points presented in the article?
- How would an expert on the topic view this article?
- Is the coverage balanced?
- What kinds of evidence does the reporter present?
- Who does the reporter quote? Experts? Administrators? General public?
- What is the basis of their information? Professional experience? Research? Observations?
- How did the reporter select the person being quoted?
- Does the reporter identify sources' expertise?
- Does the reporter document the level of expertise?

Use the above questions to begin your reading and develop a discriminating attitude. A deeper understanding of your subject will help you further evaluate the articles you are reading.

Reading Textbooks. Textbooks provide a range of information on specific subjects. Most textbooks encourage students to progressively build an understanding of the topic. Introductory texts provide overviews of general fields;

advanced textbooks focus on more narrowly defined topics while still presenting a wide range of information within that specialized area.

In literature reviews, textbooks can help you understand the general field, specialized topics within a field, and historical perspective on the field. Textbooks can also provide sources of further information, but keep in mind that it seldom will be the most recent information on a topic. Read textbooks for a general introduction and use the following questions to guide your reading.

- Why are you using this text? A professor's recommendation? An expert's recommendation?
- What's the source of the information presented? The author's expertise? Other professionals' expertise?
- What is the author's expertise?
- Does the author have additional expertise in a subfield of the topic?
- Does the text present general knowledge of the field?
- Does the text cite other sources?
- Does the text devote a chapter to the topic of your interest? A section?
- What depth of coverage is presented within that section?
- What are the primary points made within that chapter or section?
- Are there related chapters to help understand the problem?

As with critical reading of other materials, use a questioning approach. Think about the author's points, the evidence presented, and the rationale provided. Does it make sense based upon what you know about the topic? As you learn more about the specific topic, your assessment may change, but let textbooks help guide your work.

Reading Specialized Books, Monographs, and Reports. In contrast to textbooks, specialized books, monographs, and reports devote lengthy discussions to more narrowly defined topics. Such sources can be extremely useful when researching a topic. Ask such questions as

- Who is the audience for the book?
- Is the publisher well-known for publishing works on the topic?
- Is the publisher a professional scientific or technical society?
- Does the book provide an in-depth discussion of a specific topic or topics?
- Is the book part of a series on the specialized topic?
- What is the author's expertise?

- What are the academic credentials of the author?
- How many years of experience does the author have in the field?
- Does the author(s) report scientific experiments or studies?
- Does the author(s) review related literature?
- Are the citations to articles in the academic or research journals?
- Are the citations to technical, scientific, or specialized reports on the subject?
- Does the introduction acknowledge reviews by well-known professionals in the field? Other authors you have read?
- What additional evidence is presented to support the points made?
- What are the primary points made?
- What limitations does the author place on the research reported?
- What recommendations does the author suggest for future investigations?

Again, let a critical, questioning approach guide your reading of the materials. Keep in mind that scientific and technical topics can be approached from different perspectives and that many authors favor a particular perspective. Finally, the literature cited may help identify additional literature for your review.

Extracting Information and Taking Notes

As you assess and evaluate the articles, books, and reports, take careful notes to help you better understand the new information, to avoid having to reread the literature later, and to extract salient references. In note taking, selecting the needed bibliographic and content information is vital so that you can credit your sources and refer to salient topics covered in the article. Begin by reviewing the style manual and publication that you will be following. A careful review of the citation styles will help you minimize last-minute trips to the library to retrieve additional information.

As a minimum, key bibliographic information for *periodicals* includes

- Author's first and last name and middle initial
- Date of publication
- Article title
- Journal/periodical name
- Volume number
- Issue number
- Page numbers

As a minimum, key bibliographic information for *books and technical reports* includes

- Author's first and last name and middle initial
- Date of the publication
- Title of publication
- Titles of chapters or sections, if specifically reviewed
- Publisher
- Place of publication (city and state)
- Edition number, if multiple editions
- Number of pages

Once you have the bibliographic information recorded, transcribe the content information you need. At a minimum, the content information should include a descriptive or informative abstract. Descriptive abstracts talk generally about the article and seldom give specific information. Informative abstracts for research articles should provide the research question, a summary of the methodology, and the key findings.

For a more thorough review of research articles, consider including notes summarizing the following:

- Theoretical perspectives or rationale influencing the project
- Research question/objectives/ purpose statement
- Methodology—sample size, if appropriate
- Methodology—sampling methodology
- Methodology—data collection techniques
- Methodology—data analysis
- Primary findings
- Secondary findings
- General observations or insights into the problem
- Conclusions
- Recommendations for future research

In your notes, paraphrase the information. Summarize it in your own words. Record information in words, phrases, sentences, or paragraphs as needed. If you quote the article, put quotation marks around the word-for-word notes. However, when reviewing another author's work, researchers in technical communication, scientific and technical fields seldom use lengthy quotes, so limit your quotations to key points. If you plan to quote a passage and you have photocopied the article, highlight the section to be quoted and note its location by page number. If you want to quote a passage from a book, flag the passage with the self-sticking tags; note the page number(s); photocopy the page(s), the title page, and the back page; then write the bibliographic information on the top of each page. If you photocopy and highlight the passage, you minimize transcribing errors and can check the quotation against the photocopy.

Be sure to note the pages from which you have drawn key information in case you want to review the sections later. Some citation styles require that you cite the specific page number(s); others do not. A few minutes invested in recording these details will save you time when you begin writing reports and articles and preparing presentations.

CREATING A BIBLIOGRAPHIC DATABASE

How you record references depends upon your prior guidance in reviewing literature and your access to a personal computer with a sophisticated word processing program and printer. You can opt to create a bibliographic database by manually recording your references or using either a word processing program or bibliographic software.

Recording Information Manually

You may record manual notes on either the traditional note cards or 8.5"x11" paper.

Conventional guidance suggests creating bibliographic and note cards on each literature citation using 3"x5" or larger note cards. The bibliographic card includes all key information needed to provide full citations following your style manual's citation style. The content note card identifies the source by providing the author's last name, the date, and a summary of the relevant information. Such notes can be handwritten or typed. This approach enables you to organize the information by arranging and rearranging the cards as needed.

As an alternative, you could use standard 8.5"x11" paper. Type the source at the top of the page. Drop down half a dozen lines and then begin typing your notes, single spacing the copy. One advantage of standard typing paper is its capacity for more information. A disadvantage is that more information on each page could hamper the reorganization of your notes. On the second and subsequent pages, put the author's last name, the date of the publication, and the page number.

Whether you use note cards or standard typing paper, keep in mind that typing is much faster than writing notes out by hand. Typing also compresses information.

Using Word Processing Programs

With a standard word processing software, you can create one or more files of notes of the literature you are reviewing. Three additional advantages are 1) you can print the files on note cards, 2) you can use the search functions to find specific information in your files, and 3) you can copy text from your literature review files into report or article files as you are drafting the manuscript.

Many software programs enable you to specify the paper size, the paper stock, and the print for specific printers. Check the technical manuals for your software and printer for details. Tractor-feed note cards in various sizes are available for letter quality and dot-matrix printers. If you have a laser printer, check your manual to see if it can print on note cards. If the printer you are using cannot handle card stock, print the cards on regular 20-lb weight paper and trim to a size you can use. If you damage them, you can easily reprint new ones.

Most word processing programs have a "search" or "find" function that you can use to search for specific letters or words. As you work through a paper or report, you can use the "search" function to find specific information, such as articles and names, in your literature review file.

As you need citations for reports and articles, simply copy them from your literature review file to your document. A word processing program that allows you to open two files at the same time and to switch between the files will let you block copy a section of text from your literature review notes, switch to the document you are writing, and retrieve or paste the text into the second document. Using such a function reduces keyboarding mistakes and speeds writing.

Using Bibliographic Software

Bibliographic software programs can handle thousands of citations and ease the task of manually managing references. Major advantages of using these programs are that the information need only be entered once and that citations can be reformatted for your specific needs. In addition, such software can handle a sizable database. For example, Reference Manager enables you to create one database of 65,000 citations; Papyrus enables users to store 1,500 references per megabyte. The programs vary widely, but their functions often include

- Advanced searching functions
- Advanced sorting functions
- Advanced importing features from bibliographic databases

- Automatic formatting for different citation styles
- Importing citation files from word processing programs

Bibliographic software includes such programs as DMS 4 Cite, Pro-Cite, Reference Manager, WP Citation, End Note, Sci-mate, and Ref-11. Versions are available for most operating systems but most require a hard disk and 640 RAM for efficient operation. In late 1993, bibliographic software prices ranged from less than $100 to more than $500, depending upon the functions and modules. Some programs interface with standard word processing programs and use the search, retrieve, and printing functions of the word processing programs. Others stand alone with their own functions.

With some programs, you can import files downloaded from CD-ROM and online databases or you can enter data using a specified format. To convert the databases from the different formats, the software uses filters that automatically convert the files. For example, Papyrus includes filters that enable users to import files from online searches on such systems as Current Contents on Diskette, Reference Update, SilverPlatter, CD-Plus Medline, and Grateful Med; the filter also can convert files from other bibliographic databases such as Pro-Cite, Ref-11, and Reference Manager.

To enter data, you open a file, select the type of reference (journal, book, chapter, thesis, etc.), and then begin entering data line by line. For example, the data-entry screens for Papyrus, a full-function reference software, are illustrated in panel 6.2. Depending upon the software, you use either the tab, cursor, or mouse to move to each line of the screen and then enter the appropriate information. Once you've entered the information, it can be saved to a file. These programs usually come with defaults, or templates, for the different kinds of bibliographic information that you will need, such as articles, books, chapters, maps, patents, theses, or quotes. Each template requests the information needed to provide full bibliographic citations according to particular styles.

Once citations are entered into the database, you can mark those that you want to include in your paper and indicate the citation style you're using, either by journal name or major style manual, such as the Council on Biology Editors, American Chemical Society, or University of Chicago. The program will automatically provide the citation style—i.e., capitalization, quotations, underlining, periods, comma, date locations, abbreviations, and other citation details.

Should you need to use a different citation style for a different article or publication, simply indicate the different style and the program will automatically reformat the citations. Such functions can save considerable time for those who write papers requiring full citations.

PANEL 6.2

PAPYRUS SAMPLE SCREEN

The two sample screens illustrate the computer screens showing how to enter information on an article and a book chapter using Version 7.0 of Papyrus, a bibliographic software produced by Research Software Design, Portland, Oregon. Once you have entered the data, Papyrus can automatically format citations to a variety of citation styles.

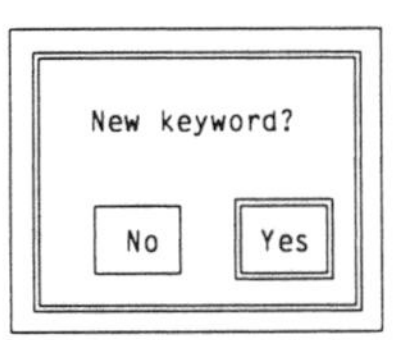

This is indeed a new keyword (as opposed to a misspelling of an existing one), so press `ENTER`. You have now added `NEUROSURGERY` to your Keyword Dictionary.

You can now enter another keyword on the next line. Type in `attitudes`, and then `surgery` as the third keyword:

```
Reference   Edit   View   Type                              ARTICLE
 Reference # | 5
   Author #1 | Ramasubramanian,SR
   Author #2 | Mull,M
   Author #3 |
        Year | 1979
       Title | Eclectic neurosurgery. Part II: Implantation of healthy
             | attitudes
     Journal | Journal of the American Medical Association
    Volume # | 280
     Page(s) | 331-348
    Abstract |
    Comments |
     Keyword | NEUROSURGERY
     Keyword | ATTITUDES
     Keyword | SURGERY
     Keyword |
```

Now save the reference. This returns us to the reference-picking question.

W38

PANEL 6.2 (continued)

PAPYRUS SAMPLE SCREEN

Let us add some keywords to another of our entries:

```
Ramasubramanian,SR (1992): Eclectic neurosurgery. In: Work Once, Publish
Forever. (Ed: Perish,N) (Career Maintenance Series, Part 27.) Academic
Press, Chicago, 201-240.
```

To find this reference we need to enter the author's name. A familiar shortcut is available to us here—press F2. Now pick Ramasubramanian,SR from the author list. After you accept Any for the year, all of this author's references will be displayed. The one we want is #6. So when PAPYRUS again asks Reference # / Author: , we can now answer "6".

By the way, there is a shortcut within a shortcut when using the F2 key. If you type the first few letters of the author's name *before* pressing F2, you don't have to wait for F2 to ask you for them.

Once you've brought up the entry screen for this reference, move down to Keyword #1 . Now add the keywords NEUROSURGERY , SURGERY and REVIEW ARTICLE . The order in which you enter these makes no difference. Remember, you can use the F2 key for keywords already known to PAPYRUS—for example, try typing "N" and then pressing F2.

```
Reference   Edit   View   Type                                CHAPTER
      Reference # | 6
        Author #1 | Ramasubramanian,SR
        Author #2 |
             Year | 1992
    Chapter Title | Eclectic neurosurgery
       Book Title | Work Once, Publish Forever
          Edition |
           Volume |
        Editor #1 | Perish,N
        Editor #2 |
     Series Title | Career Maintenance Series
  Place in Series | Part 27
        Publisher | Academic Pres
          City of |
      Publication | Chicago
          Page(s) | 201-240
         Abstract |
         Comments |
          Keyword | NEUROSURGERY
          Keyword | SURGERY
          Keyword | REVIEW ARTICLE
          Keyword |
```

Chapter 6: Modifying References W39

Selecting a Bibliographic Software

To select a bibliographic software, 1) check with other scientists and librarians to see which systems they are using, 2) conduct a search of current articles reviewing bibliographic software, and 3) obtain demo disks and manuals to review.

Major research libraries often have a staff member who follows the development of personal bibliographic software. For example, Barbara Burke, Colorado State University, provides one or more faculty workshops annually on using bibliographic software. She also knows the faculty on the CSU campus who are using bibliographic software programs. On some campuses software user groups form to share their experience and expertise in using different software programs. An inquiry to the library, computer support staff, or others will often identify experienced users to whom you may turn for assistance.

Consider conducting a search for current articles that report software reviews. For example, an online search of the Magazine Index on CARL in early December 1993 identified two recent comparative reviews (Rabinowitz 1993; Neal 1993) and one product review (Blumenthal and Gilad 1993). Such articles provide insights and comparisons that may help the novice user narrow his or her choices.

Giving the software a test drive provides insight into the ease of use, compatibility, and potential value for personal uses. Such test drives give you a full feel of the software and let you identify its strengths and weaknesses.

Appendix B provides a summary of bibliographic software, their respective features, and company addresses.

CONSIDERING ETHICAL AND LEGAL ISSUES

From a reference use perspective, you should consider ethical issues surrounding plagiarism, and legal issues surrounding copyright. Both plagiarism and violation of copyright laws are viewed seriously in professional circles. Unfortunately, problems do crop up from time to time.

Plagiarism

Plagiarism, or intellectual thievery, consists of passing another person's work off as your own without giving credit to that person. The work may be a passage, information, or ideas unique to the originator. While two people can have the same ideas, most plagiarism charges center around the similarity and extent of the narrative of passages.

At times plagiarism is inadvertent, as in forgetting to give credit to the original sources. At other times, plagiarism is conscious; individuals deliberately borrow passages, information, ideas, or whole documents from another person without giving credit (Zimmerman and Clark 1987).

Discovered plagiarism can result in a tarnished reputation or forced resignation or termination from a job. In 1987, Senator Joseph Biden's bid for the presidential nomination from the Democratic Party ended when reporters discovered he failed to credit the source of the narrative of some speeches (Kaus et al. 1987). In some cases, plagiarism may go undetected, only to surface many years later. For example, medical researcher Elias A. K. Alsabti's career ended after scientists discovered that he had plagiarized some 60 scientific articles in medical journals (Zimmerman and Clark, 1987; Broad and Wade 1982).

To avoid plagiarism, do your own work and give full credit when you use materials from others. If you do not understand how to give proper credit, study how other authors in your field use direct and indirect quotes and provide citations (Zimmerman and Clark 1987).

A Closer Look at the Copyright Law

The U.S. Copyright Law of 1976, Public Law 94-553, gave legal rights to the creators of intellectual works such as books and other printed publications, lectures, maps, illustrations, photographs, and other tangible forms of expression.

Although the 1976 Copyright Law clarified many issues surrounding older copyright laws, it still left many issues open to the interpretation of the courts. For example, U.S. District Judge William Gray provided a split ruling in 1989 that 1) gave copyright protection to the instructions embedded in microchips, and 2) ruled that a Japanese company legally copied microchip designs of a U.S. manufacturer (*Coloradoan* 1989). The second part of Gray's ruling centered on whether the company had included the copyright mark, ©, on its chips.

Computer software and hardware companies spend thousands of dollars in lawsuits, counter suits, and appeals over potential copyright infringements. Authors and publishers are likewise involved in suits in cases of potential copyright infringement, as the court decisions interpret the specifics of the Copyright Law.

At the heart of the Copyright Law is the form of the expression. Consider the numerous how-to books on using a particular software. Do they infringe on the software manufacturer's copyright of the original manual for the software? Usually not. While the underlying ideas may be the same, the form of expression—writing, organization, structure, and layout—is usually different.

In general, the products of your creative endeavors—articles, reports, books, illustrations, visuals, videos, films, and photographs—are protected under the 1976 Copyright Law. The law says you are entitled to the monetary rewards of your efforts. Exceptions, however, do exist. If you work for a company, your writings usually belong to your employer under the "work for hire" provision of the 1976 Copyright Law unless your job contract provisions specify otherwise. Simply, if the company is paying you to write, the company owns what you produce. It's much like the person working on an assembly line. The company pays for the individual's efforts, and thus it is entitled to the profits from that investment. In contrast, works produced by government employees are not protected under the Copyright Law (Kleinman 1978). Government publications usually, but not always, belong to the public because taxpayers dollars funded their production.

When you or your company publish a work, you can better protect your legal rights if you place the copyright symbol, ©, and date on your work and register it with the Library of Congress Copyright Office to document yourself as the originator of your work. For details on registering materials, call the Library of Congress Copyright Office in Washington, DC, and request the copyright forms and registration guidelines.

You might wonder how you can use works protected under the Copyright Law. The law provides for the provision of "fair use" and permissions. Fair use generally means that you can use limited portions of a work for criticism, comment, reporting, teaching, and research without permission (Kleinman 1978). Thus, you can photocopy an article from a periodical for scholarly use, but you cannot photocopy an entire book (Zimmerman and Clark 1987). Further, although students have not been sued for using copyrighted materials in term papers (Robert Dreschel 1993, personal communication), you should provide full and proper citations to avoid plagiarism. For publications being produced for profit, publishers usually provide fair use guidelines for authors. Generally, the guidelines specify that 200 to 300 words, in whole or parts, from a major book is permissible, but authors must seek permission to use drawings, photographs, tables, or illustrations from copyrighted publications. The courts also continue to interpret the "fair use" provision.

To ensure that you have not violated someone's copyright, seek permission to use their materials. Write the individual or publisher, identify the original publication, specify the pages and nature of the material that you would like to use, and how you intend to use the materials. Some publishers may give you free rights to use the materials; others may ask that you pay a permission fee. Publishers also specify how you should acknowledge their permission to use their copyrighted materials, and the duration of the permission.

REFERENCES

Baraslou, L. W. 1992. *Cognitive psychology: An overview for cognitive scientists*. Hillsdale, NJ: Erlbaum.

Blumenthal, E. Z., and R. Gilad. 1993. Storing a bibliographic data base on your PC: A review of reference-manager software. *New England Journal of Medicine* 329(4): 283–84.

Broad, W., and N. Wade. 1982. *Betrayers of the truth: Fraud and deceit in the halls of science*. New York: Simon & Schuster.

Coloradoan. 9 February 1989. Judge extends computer copyright. Money section, p. 1.

Huckin, T. 1983. A cognitive approach to readability. In *New essays in technical and scientific communication: Research, theory, practice*, ed. P. V. Anderson, R. J. Brockman, and C. R. Miller. Farmingdale, NY: Baywood.

Kaus, M., E. Clifton, H. Fineman, and J. McCormick. 1987. Biden's belly flop. *Newsweek*, 28 September 1987. 110(2): 23–24.

Kleinman, J. M. 1978. The Copyright Law of 1976. *Technical Communication* 25(1): 11–13.

Neal, P. R. 1993. Personal bibliographic software programs: a comparative review. *BioScience* 43(1): 44–50.

Nielsen, J. 1990. *Hypertext and hypermedia*. Boston: Academic Press/Harcourt Brace Jovanovich.

Osherson, D. N., S.M. Kosslyn, and J. M. Hollerbach, eds. 1990. *Visual cognition and action*. Vol. 2 of *An invitation to cognitive science*. Cambridge: MIT Press.

Osherson, D. N., and H. Lasnik, eds. 1990. *Language*. Vol. 1 of *An invitation to cognitive science*. Cambridge: MIT Press.

Osherson, D. N., and E. E. Smith, eds. 1990. *Thinking*. Vol. 3 of *An invitation to cognitive science*. Cambridge: MIT Press.

Pugh, A. K. 1978. *Silent reading: An introduction to its study and teaching*. London: Heinemann.

Rabinowitz, R. 1993. Point of reference. *PC Magazine* 12(17): 269–79. (October 12, 1993).

Zimmerman, D. E., and D. G. Clark. 1987. *The Random House guide to technical and scientific communication*. New York: Random House.

Planning and Conducting Interviews and Surveys

CHAPTER

Interviewing

Various kinds of interviewing occur daily. This chapter focuses specifically on formal, information-gathering interviews, in which you individually seek information from another person about a specific topic. Handled properly, the informational interview is an invaluable information-gathering strategy. It provides you the opportunity to learn the observations, opinions, and ideas of people closely associated with a topic.

For example, technical communicators often interview engineers, scientists, and technical specialists when collecting background information for computer manuals, instructions, magazine articles, and reports, or for videotapes, slide sets, and multimedia presentations. Technical communicators, scientists, and engineers often interview members of their intended audiences to understand them better and thereby improve communication.

This chapter explores the dimensions of the interviewing process (fig. 7.1) and provides guidance on executing more effective interviews.

INTERVIEWING FOR INFORMATION GATHERING

In its most simple form, informational interviewing entails asking questions of another person, but that is just one step in a series. First, you identify a need for specific information, develop your questions, identify a person with expertise and interview that person, and finally, you assess the information you gathered.

From a social science perspective, the informational interview is a social interaction between you and another person in which you ask for information and opinions about a specified topic. Through a series of questions, over a brief period of time, you have the opportunity to capitalize on the interviewee's background, expertise, and experiences.

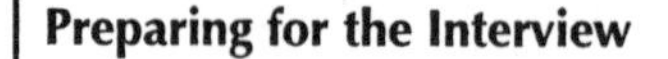

Preparing for the Interview

a. Backgrounding yourself
b. Preparing your questions
c. Identifying and locating interviewees
d. Setting up the interview
e. Planning for note taking

Conducting the Interview

a. Arriving 5 minutes early
b. Beginning the interview softly
c. Moving quickly to key questions
d. Probing as needed
e. Recapping key information
f. Arranging for a call back
g. Closing on time

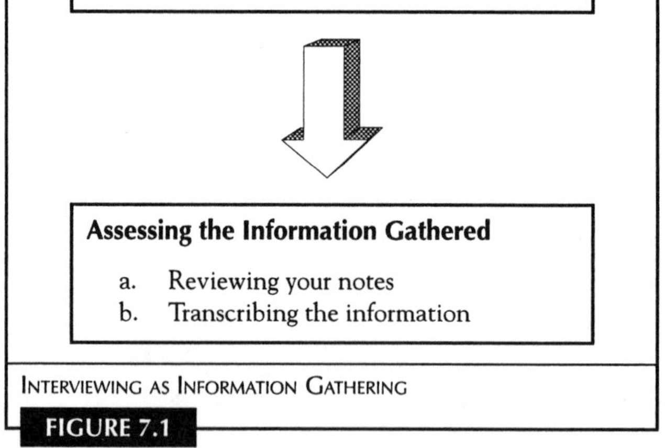

Assessing the Information Gathered

a. Reviewing your notes
b. Transcribing the information

INTERVIEWING AS INFORMATION GATHERING

FIGURE 7.1

You can take lessons in interviewing from communication science and journalism. When providing guidance for journalism majors, Steve Chaffee, a leading communication scientist who chairs Stanford's communication department, divided potential interviewees into regular people, experts, and potentates (Chaffee 1975). Any one individual can fill any one of the three roles, depending upon the topic. For example, a medical doctor is an expert on medicine, but a regular person on highway engineering. Potentates, or administrators and supervisors, may be experts in electrical engineering, but not in accounting.

Experts are subject-matter specialists on a particular topic. They understand the terminology and conceptual foundations of a specific topic, are familiar with the literature in their field, and usually have colleagues who are also experts on the topic. Experts may feel comfortable being interviewed since they are used to discussing their field with colleagues, usually in give and take sessions, and may have been interviewed previously.

Potentates oversee an organization, a department, or a section within an organization. Although some may be subject matter specialists, their primary duties involve the overall function and running of the unit they administrate. They may or may not be experts in the topic area, but their work focuses on keeping the unit operating, handling personnel matters, concentrating on budgets, and other administrative duties. Although potentates may tend to take control of interviews and avoid answering specific questions, they may provide an overview of the subject.

Regular people may have had some experiences with a topic, but they lack professional expertise. Many have not been interviewed previously and may feel uneasy. Regular people's reactions can range from being quite talkative to being very reserved.

PREPARING FOR THE INTERVIEW

The key to good interviewing is a process approach that begins with gathering enough knowledge about the topic to develop questions that elicit helpful information.

Backgrounding Yourself

When interviewing experts you do not want to ask for simple information that you could learn from a careful literature review. Such questions waste the expert's time and signal that you have not bothered to background yourself.

For example, suppose you were investigating the impact of computers on repetitive stress injuries such as tendinitis and carpel tunnel syndrome, two common injuries from heavy computer use. You do not want to waste a researcher's time learning how the injuries occur: you should already know that. Instead, you want your questions to probe the researcher's understanding of the topic and how the researcher's study relates to repetitive stress injuries from computer use.

To conduct an adequate interview, develop a working knowledge of your topic. Begin by asking yourself what you already know about the topic and what you need to know about the topic to ask good questions. To answer these questions, try a brain-storming approach in which you begin listing your responses without thinking or evaluating them. Develop a questioning attitude. Questions that can help you identify what you know about the topic include

- Who?
- What?
- When?
- Where?
- Why?
- How?

These preliminary questions should help you generate other questions. You can ask yourself, "What do I know about the topic? Do I know any researchers on the topic? Experts on the topic? Do I know when the topic emerged as a major issue? Where is the topic more important? Why is it of importance to me? How does it work?"

Study these questions for holes in your subject matter background and to identify the areas in which you need additional information. For example, you can ask, "Who are the local experts on topic X? Who are the regional experts? National experts? International experts? Who has published heavily on my topic? What are their backgrounds? What makes them experts?" After you have answered these questions you are ready to formulate the specific questions that will make your informational interview more productive.

Preparing Your Questions

Developing your questions centers around what you need to know from the interviewee, how many and what kinds of questions you should ask, and in what order you should present the questions.

First, you want to concentrate on information that you cannot gather from other sources. Consider questions designed to answer

- What do I know about the topic from person X?
- What can person X provide that I can't obtain from other people?
- What can person X provide that I can't obtain from other sources?

Second, the number of questions depends upon the topic and the time allowed for the interview. Consider having 10 to 15 key open-ended questions to guide a 30-minute interview. The interviewee could easily spend two to three minutes on each response. If you need to probe a response, the time per question rises quickly. Keep in mind that it is better to have too many questions than too few. By developing a longer list of questions, you will find yourself well-prepared and may also discover that you have follow-up questions to help guide your probing.

Third, most personal interviews produce more useful information if you use open-ended questions, or those that cannot be answered with a "yes," "no," or similar response. Your questions should engage the interviewee and encourage a free discussion about the topic. Open-ended questions generally provide richer responses than closed-ended questions—if the interviewee does find not them threatening. For a further discussion of developing questions, see chapter 11.

Fourth, carefully consider the order of your questions. When dealing with potentially threatening topics or controversial issues, consider beginning with a few general, nonthreatening questions before asking the key questions. Also, order your key questions so that you ask the more important ones first, lest you run out of time or the respondent becomes hostile or nontalkative. In that way, you at least learn the most important information.

Identifying and Locating Interviewees

As you begin to develop a familiarity with a topic, keep track of the names of individuals who might provide useful interviews. Consider these strategies:

- Reviewing literature
- Checking professional directories
- Reviewing university/college directories
- Looking through telephone directories
- Calling directory assistance
- Reviewing state and federal government directories
- Monitoring your media use

As you work through literature on your topic, note the authors and researchers mentioned in different articles, books, chapters, and reports. Many articles carry biographical sketches of authors and identify their research institutions. Note departments for which the individuals work. Check if your local universities, colleges, or government agencies have similar departments. If so, check to see if anyone in those respective departments has similar expertise.

A review of professional directories may be helpful. Some directories list professionals' telephone numbers and mailing addresses. Other directories supply areas of professional expertise or provide professional biographical sketches that will enable to you to evaluate their expertise.

A review of university and college directories may help you. In fact, public relations or public affairs offices often publish a resource directory centered around faculty expertise. Investigating a particular area could lead you to one or more researchers with such expertise. If the university or college does not have such a directory, call the public relations office to help you identify experts on the subject.

Many areas now have multiple telephone directories and most include a section for government agencies as well as universities and colleges. Always keep in mind that directory assistance is available to help you identify specific telephone numbers. Telephone operators and receptionists at large organizations can help you locate experts on their staff.

Consider checking state and federal government agency directories. Many states have directories, and the federal government has a variety of local, regional, and national directories, depending upon the respective discipline. Many government agencies have departments that cover a wide range of topics; offices in different agencies may cover the same topic.

Finally, try to spot information as you read newspapers, magazines, and newsletters and as you watch television shows and listen to radio programs.

Once you target potential interviewees, assess whether or not they may be helpful to you. Let the following questions guide your decision.

- What are the individual's qualifications?
- Why would you want to interview them?
- What kinds of information can they provide?
- What makes this individual an expert?
- What is the individual's academic training?
- What is the individual's professional experience?
- What is the individual's research expertise?
- How knowledgeable might the person be on the topic?
- What kind of information can this individual provide?

In some cases you can answer the questions based on your literature review or the review of biographical information in professional directories. In other cases, the school or agency's public relations office can give you such background information. You may want to call a potential interviewee, explain your project, and then ask a few brief questions to ascertain their expertise on the subject.

Be sure to consider what kind of information an individual will give. Will the information be based on facts, opinions, or personal observations? You may want to consider the bases of their facts, opinions, and personal observations. Ask yourself, "How did they come to know what they know?"

Setting up the Interview

Once you have identified the individual(s) you need to interview, you will need to set up an appointment for the interview. Before you do, prepare a brief explanation about the project, why you need to interview the particular person, an estimate of how long the interview will take, and how you will be using the information elicited from the interview.

Keep in mind that some people are very busy and difficult to contact. With electronic voice messaging services, you can leave messages, but their extremely busy schedule may preclude their responding to calls quickly. Additionally you may find yourself playing telephone tag—you leave a message, your call is returned with a message for you to call back, ad infinitum. With some people, you can leave a message with a real person—secretary or manager—if needed. If you find yourself playing telephone tag, it may help to do so.

After reaching the person you wish to interview, explain your need for the interview, and if the individual agrees, set a time and location. Let the interviewee suggest a convenient time and location. If possible, set the interview at a site where you will not be disturbed or the disruption can be kept to a minimum, such as offices where you can close the door and turn off the telephone. Hotel lobbies, coffee shops, restaurants, and living rooms are problematic because of noise, music, conversations, or other distractions.

Once you have set the location, ask for detailed directions. Do not assume that you know how to get there. If you will be traveling some distance, ask interviewees if they know how long the trip will take. Also check on parking; you do not want to be late due to no nearby parking. In planning your travel, try to arrive at least 5-10 minutes early so that you do not keep your interviewee waiting.

Planning for Note Taking

When planning for an interview, consider carefully how you will take notes. At a minimum you should take notes with pen and pad. (See panel 7.1 for suggestions.) Tape recorders, with their small size and easy handling, seem to be a logical alternative to note taking, but using a tape recorder has its disadvantages as well as advantages.

Advantages include an exact recording of comments from individuals and the recorder's unobtrusiveness in the interview setting. Disadvantages include the potential impact on the interview, time required to transcribe the passages, and possible equipment failure. Some interviewees find tape recording threatening and either will not allow tape recording or will not talk as freely as they would have otherwise. You should be aware that transcribing a tape recording can easily run two to four times as long as the interview itself. Such lengthy transcriptions can be useful for detailed analyses, but most profes-

PANEL 7.1

Taking Notes—The Key to Good Records

Taking good notes requires 1)MATERIALS FOR NOTE TAKING, 2) DEVELOPING A NOTE-TAKING TECHNIQUE, AND 3) REVIEWING YOUR NOTES.

While any good notepad will do, take a lesson from newspaper and magazine reporters, and use reporters' notebooks. These notebooks are 4"x 8", spiral-bound at the top, with a hard cardboard backing, much like a stenographer's notepad. The size fits nicely in your hand, the backing is firm enough to allow you to write easily and quickly, and the spiral binding keeps the pages in order. If you do not use reporter's notepads, consider using a notepad binder or clipboard for a firm backing. Whatever notepad you use, make sure you have enough paper for your notes during the interview. You don't want to run out of paper.

Also be sure to have two or three fresh pens so that you will not run out of ink. Avoid pencils. The lead breaks, and the room's light could make reviewing penciled notes during the interview difficult.

When you take notes, take another tip from journalists—don't try to capture every word on paper. Be selective. Record key ideas, facts, figures, dates, and dollar amounts. When the interviewee says something worth quoting, speed your notetaking by jotting down the only the key words. Drop *a, an, the,* and other short words that are not pertinent to the quote.

If the interviewee is talking too fast, and has made a quotable point, ask for a pause while you complete your notes. Then repeat the quote back, saying, for example, "I understand that you're saying . . ." as a preface. This gives the interviewee a chance to help you correct your notes, should they be in error.

Leave about one-third of the interview time to review your notes, to verify what you've recorded in your notes, and to ask any additional questions that you may have. If you are interviewing an individual on a controversial topic, be sure to check your notes so that you do not misunderstand the interviewee's points. Immediately after the interview, carefully review your notes and complete any words and notes as needed. Your mind is freshest at this point, and you can often add additional helpful details that you may forget in a day or two.

sionals end up using only a few quotes from a transcription. If you are considering using a tape recorder, carefully review panel 7.2.

CONDUCTING THE INTERVIEW

Use a process approach when conducting the interview, as illustrated in figure 7.1.

PANEL 7.2

HOW TO USE A TAPE RECORDER

If you want to use a tape recorder, avoid potential disasters by taking notes as suggested in panel 7.1. If you are buying a tape recorder for interviews, buy a unit with a counter and sound level indicator so that you'll know if the unit isn't working properly.

Prior to the interview, check to make sure you have

A working tape recorder
New or fully charged batteries
Back-up batteries
Back-up tapes, preferably 60 or 90 minutes

Before the interview, be sure to ask permission to use the recorder. Some interviewees will refuse and others will be resistant, in which case you perhaps should reconsider. When you begin the interview, turn the tape recorder on and place it between you and the interviewee. As you begin asking questions and the interviewee begins answering, check the recording level to make sure the unit is working properly. Throughout the interview, watch the tape counter and be ready to change tapes at an appropriate time. Finally, always take backup notes with pen and paper, and review your notes prior to the interview's end, as suggested in panel 7.1.

Arriving 5 Minutes Early

Plan to arrive about 5 minutes early and check in with the secretary or receptionist. If the interviewee does not have a secretary and is busy, wait politely until the appointed time, and then knock to signify that you have arrived and you are ready to begin. Most interviewees will then promptly invite you into their offices.

While you are waiting, double-check the spelling of names on doors, and write down the secretary or receptionist's name, if present. Review your key questions while you wait. This will help you appear more prepared, and almost spontaneous, as you work your way through the questions.

Beginning the Interview Softly

To obtain cooperation for your interviews, attempt to develop a good working relationship with the interviewee. Skilled interviewers often begin by noting something of importance in the office and generating light conversation. You may then briefly review your project and the purpose of the interview.

A soft opening can also involve emphasizing the value that your project might be to the interviewee, or the benefit their cooperation might have to their organization, themselves, or society. Simply, frame your need for information to reflect how the interviewee might see the advantage of taking time to be interviewed. Once you have established that working relationship, begin asking your questions.

Moving Quickly into the Key Questions

In most cases, you can begin with questions that establish the general focus of your interview and give the interviewee a clear idea of what you are doing. You should then move quickly to your key questions.

Try to structure your interview so that you can ask all of your key, or most important, questions about two-thirds of the way into the interview. By coming quickly to the point, you allow yourself time to probe if needed to clarify issues. Also, such a quick beginning increases the chances that you will obtain the key information should the interviewee need to close the interview before you have had the opportunity to ask all of your questions. In this way, you will have obtained at least the key information that you need. This could be especially valuable in cases where you are dealing with an extremely busy individual, a controversial topic, or an individual with a strong viewpoints.

Probing When Needed

During an interview, you may find that you did not understand the response, or that the interviewee digressed, brought up a new and relevant point, or did not answer your questions. At that point, you will need to probe.

You can probe by asking for clarification, an elaboration, a further explanation, or in some cases, by restating the question. You can ask for a clarification by saying, "That's an interesting explanation. Could you help me better understand . . . by" or "Please clarify what you mean by. . ."

You can ask for an elaboration or further explanation by saying, "I understand. . . Could you elaborate on . . . ?"

Alternatively, you could restate the question. You could ask, "Let me recast the question in a different way, . . ." (and then ask the same question again using a different approach).

Keep in mind as you work through the interview that you are interested in obtaining key information. Probing for more information, clarification, and elaboration may be the key to gaining the information you require.

Recapping Key Information

Even though you are taking notes throughout the interview, always reiterate the interviewee's answers to specific questions to make sure that you have not misunderstood. You can summarize the information by simply saying, " I understand that you're saying . . ." and then reviewing the key points.

When you quote a person, summarize the information and review what they have said. You might say, "Let me review what I understand that you have said," and then provide a succinct reiteration.

Leave time in the interview to review, summarize, or recap all of the points made. By recapping the key points, you ensure both that you understand the information and that the interviewees have the opportunity to clarify their responses. At times, individuals will err in responding to your questions; recapping gives them an opportunity to catch such errors.

Arranging for a Call Back

As you close the interview, ask the interviewee if you can call them back should you need to clarify points or seek additional information. Your approach should reflect your desire to carefully and accurately reflect what you were told. Few respondents will refuse such a request.

Closing on Time

As you work through the interview, watch the time so that you do not run too long. Again, remember that the people you interview are extremely busy and expect a business-like approach from you.

ASSESSING GATHERED INFORMATION

Develop a standard process of reviewing your notes and transcribing the information. Set aside time immediately after your interview to complete your post-interview activities.

Reviewing Your Notes

Immediately after the interview, take 15 to 30 minutes to carefully review your notes and clean them up as needed. You may discover partially spelled words, incomplete information, unique abbreviations, or failure to note some items. By working through your notes immediately, you will often recall information that you would forget after a few days or weeks. Memory fades quickly, and details are lost. Reviewing notes immediately after the interview minimizes your memory loss.

As you review the notes, you will find that you can elaborate on some points and add missing information such as interesting and useful details that you did not include while taking notes. You will find that you can enhance the quality of your notes and the information with an immediate, post-interview review session.

Transcribing the Information

You can transcribe notes from an interview either by creating a full transcription of the interview, including every note, in detail, with all of your elaborations, or by typing a summary of the key information or points made during the interview.

The full transcript approach works well if you later may need to review the entire interview or do an analysis of the information gathered. Such transcriptions are used in the areas of cognitive psychology or problem solving, usability studies of technical communications, and ethnographic research. In all cases, a full transcription provides a complete historical record.

The summary approach works well if you only need key information from the interviewee, as in this case, when you need a few pieces of information that answer a specific question. Journalists usually follow this approach for transcribing as they only need the information for an article, or series of articles, that they are writing.

REFERENCE

Chaffee, S. 1975. The interview as a reporting tool. *Gathering and Writing the News: Selected Readings*, ed. R. L. Moore, R. R. Cole, D. L. Shaw, and L. E. Mullins. Washington, DC: College Press.

CHAPTER

Conducting Group Interviews

As an alternative to individual interviews, group interviews provide the advantage of capitalizing on the collective experience, background, and expertise of a group of individuals. More and more technical communicators, scientists, and engineers are using group interviewing processes to better understand how their products are used, how potential publics understand technical and scientific issues, and how diverse groups stand on specific issues.

To give you a foundation for group interviewing, this chapter explains focus group interviewing and the Nominal Group Technique, and also provides guidance for deciding which is most appropriate.

EXPLORING FOCUS GROUP INTERVIEWS

In the focus group interview process you bring together 6 to 12 purposively selected people and ask them a series of questions in a structured manner. As the moderator, you keep the group focused on the topic so that you can learn as much as possible from the group. Meanwhile, an assistant records responses and notes group reactions and comments to your questions. After the session, you and the assistant summarize the findings. The information gathered will be qualitative in nature—general, descriptive data. Focus group interviews provide 1) a good exploratory technique, and 2) a clarification technique for information gained from other sources.

Often used in marketing, advertising, and occasionally in social science research, focus groups have wide applicability. You can use focus groups to gather information; explore issues surrounding a problem; generate ideas to solve a problem; assess reactions to topics; gain further insights into a question, issue, or proposed solution; or evaluate prototype products, advertisements, brochures, or other communication.

For example, the Cancer Communication Office of the National Cancer Institute, Department of Health, Education, and Welfare uses focus groups to obtain insights into target audiences' perceptions, beliefs, and language when developing brochures, leaflets, or public service announcements (Romano 1982). Hewlett Packard's technical communicators and marketing staff use focus groups to develop a better understanding of the users of their printed and online manuals and instructions and their computers, scientific instruments, scanners, and other products.

The following discussion—based conceptually on Wimmer and Dominick (1991), Higginbotham and Cox (1979), Romano (1982), and Krueger (1988)—highlights the steps in the focus group process. These steps include

- Identifying the problem
- Determining the number of focus group sessions
- Selecting the participants
- Preparing for the focus group session
- Conducting the session
- Analyzing the session
- Preparing the report

Identifying the Problem

Begin by carefully developing a detailed understanding of the problem to which your focus groups will respond. What questions would you like the group to answer? Write them down, and set them aside for a day or two while you speculate on the possible responses. How might the focus group participants respond to your questions? Will the responses to the questions give you the information you need? If not, revise the questions. Once you have the questions clearly established, prepare a written guide that includes the questions and notes you will follow during the focus group session. As you prepare the guide, keep in mind that you can influence the group's direction with your comments, so carefully re-examine them. The questions will guide you as you conduct the session and help keep the group on the subject.

Determining the Number of Focus Group Sessions

Many professionals generally recommend at least two focus group sessions with participants having the same characteristics. Other professionals recommend four or more focus group sessions with similar participants. The number depends on the similarities of responses and the available resources for the multiple sessions.

Selecting the Participants

For each focus group session, you will need between 6 and 12 participants. Select the participants purposively. You will want individuals with knowledge or relevant opinions on your subject. Sometimes these individuals should have similar characteristics and at other times, different characteristics.

When you ask people to participate in a focus group, give them a general explanation of what you will be investigating, but do not explain the specifics because you may predispose them to a viewpoint. Of course, tell them the time, date, and location for the focus group session. Depending upon your questions, sessions can run up to two hours, but most run about an hour.

Preparing for the Focus Group Session

You will need a comfortable room for each session. Some researchers find that an informal, living-room like setting produces better results; other researchers believe that participants should sit around a table to facilitate discussions. All researchers agree that you need a neutral, isolated site to minimize distractions or interruptions. Make sure that each site has enough chairs for everyone and is clean, comfortable, well-lit, and quiet.

To provide a record, either tape record the session or videotape each group's discussion. Charge or replace the batteries in the equipment before the sessions and always have spare batteries available. Likewise, have enough blank tapes available for three hours or so. Keep extra tapes handy, should one break, become jammed, or malfunction. The day before the session, and again an hour or so just before, check to make sure the equipment works and that you have extra batteries and tapes. Label the tapes with the date, session, time, and place.

Be sure to develop session guidelines, a working outline of your plans, and questions for conducting each focus group session. Find someone to serve as an assistant to run the equipment and keep notes as you conduct the session. Provide the assistant with note pads and pens along with a copy of your session guidelines. To be sure your assistant can run the recording equipment, set up a practice time for such tasks as turning the equipment on, changing batteries, adjusting sound levels, and other equipment functions.

When conducting focus groups, professionals often provide refreshments, meals, token gifts, or honoraria as an incentive to take part in the study. When conducting focus groups with students, researchers usually provide pizza and soft drinks or coupons for sandwiches at area restaurants.

Finally, select a time for the session that is convenient for the participants. You may want to avoid holidays and weekends. If you are working with students, avoid scheduling focus groups near examination periods and other heavy demand periods. If you schedule focus groups during such periods, some participants who initially agree to participate may change their minds at the last minute.

Conducting the Session

Your careful preparation produces the best focus group sessions. Keep in mind that the room arrangements influence how much people participate. Wells (1974) points out that people who have direct eye contact, those who are directly across from the moderator, often participate more than individuals with indirect eye contact.

Open the focus group session with a general explanation of what you will be doing. Encourage everyone to develop their ideas since each person can help you gain a better understanding of your topic.

Focus group experts recommend that you keep the session free-flowing, listen carefully, respect respondents' responses, and control undue individual influence over the group. Controlling the group requires moving onto another question when several people begin echoing each others' responses. Answers like, "I agree," "Yes," and "No," do not give you the kind of information you need. In response, you can ask participants to explain why they agree, and then probe, asking additional questions to develop a deeper understanding.

A major drawback of focus groups are those individuals who try to dominate the session. They may think they are experts on all topics. If possible, avoid having just one or two participants who consider themselves experts. You may, of course, have focus groups in which all the participants are experts on the topic. If you discover a dominating person during a session, Wells (1974) recommends that you avoid eye contact, cut the person off in mid sentence, or solicit others for their opinions.

Depending upon your topic, you may need to differentiate participants' knowledge from their opinion. When you recognize a response as expressing an opinion, ask for clarification and ascertain the individual's reasoning by posing probing questions. Try hard to understand what point each participant is making and the reasons for making that point.

Develop active listening skills. First, try to clarify what the participant is saying by rephrasing the response so that the participant has the opportunity to clarify your understanding of what has been said. Probe when necessary to clarify the responses, but keep the session moving along. Active listening strives to get at a deeper understanding of the topic being discussed. As you ask for clarification, give "I messages." An "I message" focuses on your inability to understand the message, not the participant's inability to communicate. Suppose you were conducting a focus group to ascertain perceptions on smoking. In a roundabout way, one respondent asserts that secondhand smoke is not a problem for nonsmokers. To better understand her perspective, you might say, "Betty, I'm not sure what you are saying. Are you saying that secondhand smoke is not harmful to others? Could you elaborate on that for me?"

Analyzing the Session

When analyzing the information collected, three strategies are useful: immediately after the focus group session, 1) prepare an impressionistic summary, 2) analyze notes, and 3) analyze the tape transcriptions (Wells 1974, Krueger 1988).

To write the impressionistic summary, sit down immediately after the focus group session, and ask yourself, "What were the key points made? How did the group react to the overall session? Did any problems emerge during the focus group session? What is my general impression of the session?" To capture your own impressions, try brainstorming. Begin jotting down your responses to questions, clarifying points, and documenting the session as you see it.

To analyze the notes, transcribe your notes and the assistant's notes, and transcribe the tapes. As you edit the written notes and transcriptions, bracket or highlight key thoughts or responses, and add marginal notes as needed. Then cut the responses apart by the bracketed items and sort those items into logical groups or topic areas. Review these groupings and topic areas and then write a report based on your overall assessment of the major areas.

For an in-depth analysis of the tapes, transcribe the tapes and then conduct a content analysis. Content analyses consist of creating a series of categories and subcategories, and then methodically counting the number of responses that fall into each category. A lengthy discussion of content analysis is beyond the scope of this book; See Babbie (1992), Wimmer and Dominick (1991), Hsia (1988), and Stempel and Westley (1989) for guidance on content analysis in general, and Berelson (1952), Gerber et al., (1969), and Krippendorff (1980) for content analysis of focus groups in particular.

After you have completed at least two focus group sessions, consider whether the responses appear to be similar. If not, then consider running additional focus group sessions. If after additional sessions a similar pattern of responses begins to emerge, you can develop a better understanding of the topic. If not, you can explore the potential reasons for those differences.

Preparing the Report

Once you have the data collected and analyzed, write a report detailing your findings. By rethinking the focus group sessions, the information gained, and your analysis of the sessions, you should develop a generalized understanding of the participants' responses to the particular questions.

EXPLORING THE NOMINAL GROUP TECHNIQUE

The nominal group technique (NGT) emerged to overcome problems associated with other group problem-solving techniques (Delbecq et al. 1975). A major weakness of many group techniques, including focus groups, is the potential of one individual dominating the conversations and strongly influencing sessions' outcomes. NGT's name reflects the fact that it is a group in name only. In other words, you receive responses from each person, individually, even though you have a group assembled. NGT has the distinct advantage that no one individual can dominate if the group leader adequately prepares for and administers the session. Further, since NGT responses can be converted to numerical data more quickly than focus group data, with frequencies, ranges, and averages for the different responses generated to a narrowly defined question, NGT data can be quantitative as well as qualitative.

The NGT provides ways of analyzing audiences, gathering information about their backgrounds, and assessing their reactions to communication products. For example, Tharp and Zimmerman (1992) used the NGT to assess student reactions to desktop publishing systems; Lloyd-Davies and Zimmerman (1992), to assess students' perceptions of the strengths and weaknesses of a university catalog; and Zimmerman (1982), to determine scientific and technical majors' perceptions of their writing problems.

The following discussion, based conceptually on Delbecq and colleagues (1975) and Moore (1987), outlines the process of preparing for and conducting NGT sessions. The process includes

- Developing questions
- Determining the number of sessions
- Selecting participants
- Preparing for the session
- Beginning the session
- Listing nominations
- Clarifying items and eliminating duplicates
- Selecting individual items
- Ranking individually selected items
- Tabulating the results

Developing Questions

NGT sessions often focus on one narrowly defined question. You can develop a NGT question on any topic you are interested in investigating, but spend time thinking about and working with the question. Slight changes in the question construction can drastically alter responses. For example, asking,

"What problems or difficulties do you face in becoming a more effective communicator?" allows for greatly diverse responses such as problems with oral presentations, one-on-one conversations, telephone discussions, and more, so think carefully about the potential responses. Prepare a response worksheet as illustrated in panel 8.1. Leave plenty of space on the page for participants to list their responses. Make 15 or so copies, so you'll have some extras should you need them.

Determining the Number of Sessions

The number of sessions depends upon the nature of the problem you are addressing, the potential pool of participants for NGT sessions, and the responses that emerge from the initial sessions. If the problem involves issues unique to a small organization with only a few participants available, you will need to limit the number of sessions. If, on the other hand, you have large pool of participants, consider running three or more sessions. If a similar pattern of responses emerges after three sessions with similar participants, consider whether further sessions would provide any additional useful information.

Selecting Participants

Usually, you will purposively select 6 to 12 participants representing the individuals who are knowledgeable of the subject you are researching. The individuals' backgrounds may influence their responses, so give careful thought to whom you select and why.

When you ask people to take part in NGT sessions, give them a succinct explanation of what you will be doing, but do not tell them the exact question or provide them with a detailed explanation of the process that could bias their responses. Be sure to tell them the time, date, location, and how to reach the room for the session.

Depending upon the group involved, you may need to provide incentives to participate in the project. If your potential participants are students in a class or employees of a company or government agency, many will be willing to participate simply to help solve the problem at hand. In other cases, if you need to assemble people as is often done for focus groups, then consider incentives to group participants as discussed previously.

Preparing for the Session

You will need a comfortable room with tables and chairs for the sessions. Make sure the room has sufficient wall space for taping a dozen or so large sheets of newsprint paper to the walls. When selecting a room, choose a quiet location with no interruptions. Plan to have the room for about two and one-half hours so you will have time for your introduction, the NGT process, and

PANEL 8.1

SAMPLE NGT WORKSHEET

Name ____________________

Student Identification Number ____________________

Class ____________________

Date ____________________

Some students have expressed concerns about their interviewing skills, while other students feel quite confident of their interviewing skills. To help us focus our class discussion on the more commonly recurring problems, we need your suggestions. For the next five minutes or so, think about the question listed below. Then briefly list your answers.

Question: What problems or difficulties do you face in becoming a more effective interviewer?

cleaning up the room. While a modified NGT can be run in less time, planning for a longer session makes the process more manageable.

Collect your supplies a couple of days in advance of the session. You will need

- 20 to 40 sheets of 14"×24" newsprint or flip chart paper
- Masking tape to hold the newsprint to the walls
- Two or three new wide-tipped felt markers
- Five 3"×5" note cards for each participant

- Pencils for each participant
- 8.5"×11" pieces of typing paper for name plates
- 8.5"×11" pad
- Felt tip pens for preparing name plates
- Worksheet with the NGT question(s)
- Small rubber bands to hold five 3"×5" inch cards together
- Small calculator with addition, subtraction, multiplication, and division functions

Next, prepare a set of notes that includes the question(s) to be asked and a working script of the NGT process you want to follow. By making these working notes, you prepare yourself for the presentation and can "rehearse" the sequence in your mind. Good preparation smooths the process.

Plan your session so you can arrive in plenty of time—an hour before, if possible—to set up for the session. Tape panels of newsprint on the front wall for writing responses to the questions. Use two sheets per panel so ink from the markers does not mark the walls, and tape all four corners of the sheets. The number of panels you need depends on how large you write, but do write large enough so everyone in the room can easily read the responses. As a minimum, tape eight panels to the wall, and have a couple of extra sheets ready should you need them. Avoid delaying the listing process during the NGT session.

Next, set up the seating in the room. Ideally, you should arrange the tables in a "U" shape and place the chairs on the outside of the "U." In that way, you can stand at the front of the group and see all members of the group, and all the participants can see each other. As the participants begin to arrive, welcome them, ask them to select a seat, and ask them to prepare a name plate by folding a sheet of typing paper in thirds, writing their name on it, and placing it on the table so you and others can read their names. As you meet them, address them by name so you can begin to associate their names and faces.

If everyone has arrived by the starting time, then begin the session. If not, wait no more than five minutes to begin. It is unrealistic to make 10–13 people wait longer than that for one or two individuals. If individuals arrive late, have them take a seat.

Beginning the Session

To begin, explain the basic process of the NGT session. Simply, it consists of responding to a specific question in writing. Each person then nominates items from his or her list, discusses the items, selects the items personally considered of most importance, and ranks those items.

Explain that each individual's ideas and responses are very important to the process and that you cannot allow discussions during the process. Specifically, you do not want group participants making value judgments or assessing the importance of items. The best way to ensure that the NGT process generates the needed information is to ensure that participants do not converse during the session. If participants begin talking among themselves, ask them to stop. As moderator, you are in control and must keep the session going.

Pass out the NGT worksheets and pencils. Read the instructions and ask participants to spend the next 5 to 10 minutes thinking about the questions and succinctly listing their responses on the worksheet. While the participants are working, review your notes in preparation for the next step. After five minutes, look to see how many participants are finished and how many are still working. After eight minutes or so, ask participants to finish their listing so the group can move on to the next step.

Listing Nominations

First, ask one of the participants to record all nominated items on the 8.5"×11" ruled pad as you list them on the newsprint. Instruct the participant not to number the items until you do so. Then start going around the room, asking each participant to nominate one item from his or her list. Ask participants to keep their responses brief enough to list on the newsprint panels. As you ask each participant to nominate an item, avoid interpreting, making comments, or suggesting a shorter presentation. As you list the item, use the participant's wording. If the wording is too long, ask the participant to shorten it. Do not number the items as you list them on the panels.

Continue going around the room, asking each participant to nominate one item per round off of her or his list. The process may continue for 5 to 10 rounds until everyone has completed nominating their items. Even after some participants have nominated all the items on their lists, check with each person as you continue going around the room asking for nominations. Sometimes people will think of additional items to list while others in the group are nominating items. Once you have completed listing all the items, ask if anyone has a final item that they would like to nominate for the listing. Altogether, this step should take 15 to 20 minutes.

Clarifying Items and Eliminating Duplicates

After you have completed listing items on the panels, read each item aloud and review the entire list. Ask the group whether everyone understands each item. If someone does not, have the group discuss the item and clarify it. Then ask the group if any of the items are duplicates. If so, ask both the group and the person who nominated the item(s) if the items can be eliminated or collapsed. If the two items mean different things to different participants, leave the items on the list; do not collapse or eliminate them. Once you have a master list, letter the items A through Z, and, if needed, AA through ZZ. Lettering the items reduces the possibility of confusion of items during the subsequent steps. Do not number the list. Allow 10 to 15 minutes for clarifying items.

Selecting Individual Items

Hand each participant five of the 3"×5" note cards, and ask each participant to review the items listed on the panels. Then say, "From the listed items, select the five that are most important to you. Write the letter and complete description of each item on its own card." Allow 5 to 10 minutes for them to select and list the items.

Ranking Individually Selected Items

Now you are ready to have participants rank the five items that they have selected. Delbecq and colleagues (1975) recommend the following dialogue.

> Now I would like you to concentrate on the five items you have selected. Place the cards on the table in front of you. After studying them, select the topic that is the most important to you. Place a number 5 in the lower right hand corner and turn the card over.
>
> Next, consider the four remaining cards. Select the least important item. Write a number 1 in the lower right-hand corner and turn the card over.
>
> Now, look at the three remaining cards. Select the most important item and place a number 4 in the lower right hand corner and turn it over.
>
> Of the two remaining cards, select the least important one and place a number 2 in the lower right hand corner and turn it over.
>
> For the remaining card, place a 3 in the lower right hand corner and turn it over.

This ends the ranking process.

Tabulating the Results

You can use either of two methods to tabulate the results, depending upon the available time and plans for subsequent NGT sessions. The preferred way is to go through the items on the newsprint panels, have participants indicate whether they selected that item, and then ask for their ranking for that item. You then list the ranking for the items and calculate the average. Work your way through each item on the list, identifying the selected items, obtaining their rankings, and calculating the averages. When you have completed the process, have the participants place a rubberband around their cards and turn in their original worksheets and cards.

As an alternative method, simply have the participants place a rubberband around their cards and turn in their worksheets and cards. You can then sort all cards by the topics selected, compile a list, add the ratings for each item selected, and calculate an average. When reporting the results of NGT, some researchers list items, others report the items and list each rating (Moore 1987), and others report the listings and quantitative data (Tharp and Zimmerman 1992). When you generate tables reporting NGT data, they should include, at a minimum, the selected items, the number of individuals selecting the items, and the average (table 8.1). Such data should help readers better understand the group's responses to the question. Other statistics could be calculated and reported, such as the percent of the group selecting the item, the range of responses, mode, median, and the standard deviations. If your data produces some surprising results, point them out, but don't dwell on them exclusively

Keep in mind that the data includes two major elements. First, the items selected by the group are the ones they believed were the more important. Second, the weights assigned give an importance value to the items. In some cases, a high percentage of the participants may select an item. In other cases, they may rate the item high in importance value.

TABLE 8.1

THIS TABLE IS AN EXAMPLE OF HOW MINIMAL DATA FROM AN NGT SESSION COULD BE PRESENTED.

Problems of journalism majors using computers during an introductory newswriting class (n=15).

Problem Category	Frequency	Mean
Software	32	3.25
Laboratory management	19	3.20
Hardward/software (could be either)	9	3.20
Training	7	3.20
Hardware	1	1.00

SELECTING THE APPROPRIATE TECHNIQUE

When considering group interviewing techniques, you need to determine which technique to use and when to use a technique. Each question involves several issues.

Which Technique Do You Use?

When faced with the task of determining whether to use a focus group or a nominal group technique to interview a group of individuals, consider

- The number of questions to be explored
- The available time
- The kind of information you need
- The expertise required

You need to consider the interplay of the four factors. If you have a number of questions to explore, a focus group can move you through the questions in a two-hour session. In contrast, NGT sessions usually concentrate on more in-depth responses to one question in a two-hour session. Focus groups provide general qualitative, or nonempirical, information; nominal group sessions can generate both qualitative and quantitative, or empirical, information.

You must also consider the kind of expertise required. Focus groups can include individuals with a wide range of expertise on your topic. For example, when the National Cancer Institute researchers use focus groups to see how the general public will respond to television public service announcements, they include focus groups with teenage cancer patients, and with their parents (NIH 1989, Romano 1982).

In contrast, NGT sessions usually require participants who are more knowledgeable on the questions to be posed, such as the manufacturing teams in industry (Henrich and Greene 1991), students in writing and editing classes (Tharp and Zimmerman 1992; Zimmerman 1982), and pharmacy practitioners (Justice and Jang 1990).

When to Use a Technique

Clearly, focus groups and NGT sessions are used in day-to-day, problem-solving activities in industry, government agencies, and educational institutions to gather information for decision making. The techniques are sometimes used by themselves as the only data-gathering technique for identifying problems, clarifying problems, and identifying potential solutions to problems.

In more evaluative and research settings, group interviews have frequently been used to provide researchers with background on topics. The researchers then develop questionnaires and conduct surveys to explore the specifics. Researchers often use group interviews of potential members of the population being studied to help them shape their understanding of the problem and to guide the development of questionnaires.

In a few cases, researchers have used focus groups after surveys to help them clarify particular issues that emerged after a careful analysis of the survey data. In such cases, group interviews may be able to answer the "whys" of responses to survey questions (Grunig 1990, 1993).

REFERENCES

Babbie, E. R. 1992. *The practice of social research*. 6th ed. Belmont, CA: Wadsworth.

Berelson, B. 1952. *Content analysis in communication research*. New York: Free Press.

Delbecq, A. L., A. H. Van de Ven, and D. H. Gustafson. 1975. *Group techniques for program planning*. Glenview, IL: Scott, Foresman.

Gerber, G., O. Holsti, K. Krippendorf, W. Paisley, and P. Stone. 1969. *The analysis of communication content*. New York: Wiley.

Grunig, L. 1990. Using focus group research in public relations. *Public Relations Review* 16(2): 36–49.

Grunig, L. 1993. Image and symbolic leadership: Using focus group research to bridge the gaps. *Public Relations Review* 5(2): 95–125.

Henrich, R. H., and T. J. Greene. 1991. Using the nominal group technique to elicit roadblocks to an MRP II implementation. *Computers and Industrial Engineering* 21(1–4):335–38.

Higginbotham, J. B., and K. Cox. 1979. *Focus group interviews: A reader*. Chicago: America Marketing Association.

Hsia, H. J. 1988. *Mass communication research*. Hillsdale, NJ: Erlbaum.

Justice, J., and R. Jang. 1990. Tapping employee insights with the Nominal Group Technique. *American Pharmacy* NS30 (10): 43–45.

Krippendorff, K. 1980. *Content analysis: An introduction to its methodology*. Newbury Park, CA: Sage.

Krueger, R. A. 1988. *Focus groups: A practical guide for applied research*. Newbury Park, CA: Sage.

Lloyd-Davies, P., and D. Zimmerman. 1992. Students' experiences using Colorado State's general catalog: Toward improving the general catalog. Technical report. Fort Collins, CO, Center for Research on Writing and Communication Technologies, Colorado State Univ.

Moore, C. M. 1987. *Group techniques for idea building*. Newbury Park, CA: Sage.

National Institutes of Health. 1989. *Making health communication programs work*. Washington, DC: National Institutes of Health, U.S. Department of Health and Human Services.

Romano, R. M. 1982. *Pretesting health communication*. NIH Publication No. 83–1493. Bethesda, MD: National Cancer Institute.

Stempel, G. H., and B. H. Westley. 1989. *Research methods in mass communication*. Englewood Cliffs, NJ: Prentice-Hall.

Tharp, M. T., and D. E. Zimmerman. 1992. Lessons learned and students' assessments of using desktop publishing in an editing class. *Technical Communication Quarterly* 1(2): 77–92.

Wells, W. D. 1974. Group interviewing. In *Handbook of marketing research*, ed. Robert Ferber. New York: McGraw-Hill.

Wimmer, R. D., and J. R. Dominick, 1991. *Mass communication research: An introduction*. Belmont, CA: Wadsworth.

Zimmerman, D. E. 1982. Identifying students' writing problems from their perspective: An exploratory study. *Proceedings of the 30th International Technical Communication Conference*. E127–30.

CHAPTER 9

Planning Surveys

As you explore the literature, you may notice that technical communicators, scientists, engineers, and technical specialists rely on a wide range of surveys for needed information.

Surveys may look easy—you merely generate a few questions, you ask a few people to respond to the questions, and then you analyze the results. Done properly, surveys provide useful data for decision making, problem solving, and scientific investigations. However, done improperly or casually, they produce inferior data, erroneous conclusions, and costly mistakes. Consider the following three examples. On first examination each looks legitimate. However, closer examination illustrates potential problems and the pitfalls of improperly handled surveys.

Example One

One summer, a natural resource manager sent college students to high mountain lakes to interview lake users. He assigned each student a dozen lakes and told them to interview lake users. The natural resource agency was planning to make its management decisions based on the survey.

Since the students received no direction on how to select or sample the individuals visiting the high mountain lakes, the haphazard sampling did not follow standard statistical sampling procedures and did not account for the potential differences in lake visitors at different times of the week or different weeks of the summer. Further, the survey did not address the likelihood that users in the summer might differ from users in the spring, fall, and winter. Finally, such haphazard sampling precludes any generalizations from the data. The resource manager could only draw conclusions about the individuals interviewed, because he had no way of knowing whether they truly represented all lake visitors. Management decisions made on such data can result in wasted tax dollars and mismanagement of natural resources.

Example Two

A state park agency surveyed registered boat users in an inland state using a mail questionnaire. The agency researcher initially reported that 50 percent of the state's 50,000 registered boats were registered to commercial marinas.

Upon initial examination of the survey results, an alert park ranger inquired, "How can the 30 commercial marinas own 25,000 of the 50,000 registered boats in our state? That doesn't make sense to me." Upon checking with the commercial marinas, the researcher found that each marina owned less than 100 boats. In all the marina-registered boats totalled less than 3,000 boats. Thus, the agency researcher had erred by 22,000 boats. How had the error crept into the survey? The alert park ranger's questioning sent the researcher back to the data looking for possible errors such as improper coding, data entry, or analysis. In this case, understanding the problem and the population being surveyed alerted a thoughtful employee to a possible error in the survey.

Example Three

A townhouse builder contracted with a commercial firm to determine which media he should use to advertise his townhouses. The survey produced a 20 percent response rate, and the data suggested that potential buyers were heavy radio users. After five years of investing $20,000 each year in one radio station's advertising with no significant home sales improvement, the builder began questioning his investment in radio advertising.

In this example, neither the commercial firm that conducted the survey nor the builder realized that nonrespondents to the initial survey may have been different from the respondents. After years of slow sales, the builder asked trained communication consultants to survey townhouse homeowners. Their survey produced a 60+ percent response rate and confirmed that townhouse homeowners do listen to the radio but less than 15 percent of them listen to any one radio station. The trained communication consultants knew of the growing body of research shows that nonrespondents to surveys are often quite different from respondents.

While surveys are deceptively easy looking and the results plausible on first examination, surveys can be fraught with pitfalls as the above examples illustrate.

Correctly executed, a survey follows a problem solving process in which you carefully articulate your problem and objectives for the survey; identify the population to be surveyed; develop a sampling strategy; draft and pretest the questionnaire; collect, code, and analyze the data; interpret the data; and report the results.

A survey requires careful attention to a multitude of details, or what is called a total design method (Fowler 1990; Fowler 1993; Dillman 1978). This approach capitalizes on decades of survey experience and research from the social sciences and usually produces a 60 to 90 percent response rate. To help you implement a careful survey, this chapter provides guidance on clarifying the problem and understanding administrative issues. Later chapters will discuss identifying and sampling the population, developing questionnaires, collecting data, and analyzing and interpreting data.

CLARIFYING THE PROBLEM

As with other problem-solving strategies, you must develop a clear statement of the problem and then ask yourself

- Why do I need a survey?
- What kind of information do I need?
- What are my objectives?
- What kind of decisions will I make based on the information gathered?

Why Do I Need a Survey?

Before beginning a survey ask, "Why do I need a survey?" At times, a thorough review of the literature and other data sources may uncover information that will negate the need for a survey. At other times, such a review may provide background information to help you develop a better questionnaire. Existing surveys can also provide background for building stronger surveys as well as provide examples of questions.

In planning a survey to explore the pollution prevention practices of small- and medium-sized Colorado businesses, Long, Zimmerman, and Boirasky (1993) conducted an online literature review using search terms such as pollution prevention, environment, attitudes, behavior, information, and public relations. Reviewing other studies helped the researchers develop a better survey instrument than they would have otherwise, one that provided more useful insights into the pollution prevention practices of businesses.

When reviewing other studies, review the overall findings. Even if the survey does not provide the exact information you need, some of the questions asked may help answer your own research questions. If so, consider whether you could use these questions in your survey. By using carefully developed questions that have been successfully used in other studies, you avoid asking questions that will not produce usable data. Keep in mind, however, that since the objectives, or purposes, of surveys vary, the questions that worked for another survey may not produce the information that you need.

In some cases, questions may not initially produce the desired results and may require several pretests. For example, when you ask most people about their family's annual income, they automatically think about annual salaries and not income from interest, rental properties, increased value of investments, or other money sources. To ascertain all income, you would need to ask additional questions designed to elicit specific information.

When reviewing existing studies, check to see if researchers report difficulties with specific questions or approaches in their respective studies. Also, check to see if they report other problems with questions that did not produce the desired results. Simply, learn from others' mistakes.

What Kind of Information Do I Need?

When you survey people, you can obtain reports of their behavior, opinions, attitudes, and observations. However, although people can often tell you how they think they behave or what they think they have done in the past, keep in mind that this is seldom accurate. People often report more prosocial behaviors than they really perform. For example, political pollsters regularly find that people report voting more than they actually vote, and during the energy crisis of the early 1970s, researchers found that most homeowners reported setting their thermostats at 68° F—the socially correct answer.

Assessing people's observations can be straightforward or tricky, depending upon the kinds of observations you are trying to assess, the time since their observations, and their observational skills. To illustrate, consider this example. A doctor, his wife, and two teenagers went on a canoe trip into the Boundary Waters of Northern Minnesota. After they had set up camp and had dinner, they retreated to their tent to play cards. Then, according to their accounts, a massive bear—more than five feet tall when standing on all four feet—raided their camp. While black bears do get big, the family's fear, their excitement, and the darkness had, in all likelihood, inflated the bear's size—few if any black bear stand five feet at the shoulders when on all fours.

Whether recalling behaviors or observations, people's memories fade, and they often remember some events in more detail than they remember others.

What Are My Objectives?

Too often, individuals untrained in survey research begin developing questions for a survey without pondering their objectives. To begin a project, ask yourself about your objectives. Write down your problem statement and then generate a series of more refined problem statements to further enhance development of specific questions.

For example, the Pollution Prevention Partnership, a group of Colorado businesses, had offered workshops, seminars, and meetings designed to help businesses prevent pollution. But few representatives of medium and small-sized businesses attended these sessions. To better understand its audiences and to prepare for an information and education program, the Pollution Prevention Partnership decided to survey these businesses and developed seven objectives to guide the development of the survey:

1. To determine small- and medium-sized business comprehension of pollution prevention as a means of solving environmental problems
2. To identify target sections that would benefit from pollution prevention technical assistance programs emphasizing the following industries: dry cleaning, automotive and fleet maintenance, printing, metal fabrication, microelectronics, biomedical, and aviation
3. To determine the extent to which pollution prevention programs have been implemented, are planned to be implemented, or have been tried and terminated
4. To identify pollution prevention opportunities for small- and medium-sized Colorado businesses
5. To identify specific barriers to implementing pollution prevention in small- and medium-sized Colorado businesses
6. To determine information and technical needs necessary to integrate pollution prevention into small- and medium-sized Colorado businesses
7. To ascertain the best means of providing small- and medium-sized businesses with pollution prevention information and technology (Pollution Prevention Partnership 1993)

These objectives provide an exceptionally solid foundation for refining further sub-objectives for question development.

What Decisions Will I Make?

When you consider a survey, ask, "What decisions will be made based on the information gained?"

The potential answers are wide-ranging. In academia, professors often use surveys as a primary tool for both basic and applied research. Technical communicators and software engineers sometimes survey customers to learn what customers think are the strengths and weaknesses of programs, manuals, and instructions; how the programs can be used; in what unique ways people use the software; and potential problems. A professor may survey a technical communications class to determine students' computer expertise to decide whether or not to develop computer-based writing assignments. A builder may survey potential home buyers to decide what kinds of homes to build as well as their sizes and desired features. To prepare for a public relations campaign on siting a toxic waste dump, a waste-management company may survey area residents to ascertain their understanding of toxic-waste handling and their attitudes toward environmental issues. The company can then use that information to determine the information it must provide the local media, community leaders, and area residents as it seeks the required permits.

In any decision making process, surveys can provide some of the information needed, but they should not be the sole source of information. In gathering information, social scientists suggest triangulation—gathering data in different ways to validate the information—such as direct observation, indirect observation, literature reviews, examination of local records, focus groups, nominal group techniques, and other methods.

UNDERSTANDING THE ADMINISTRATIVE ISSUES

When you undertake a survey, consider

- Which survey method I should use?
- How much will my survey cost?
- Gaining human subjects committee approval

Which Survey Method Should I Use?

Surveys can be categorized as self-administered, mail questionnaire, telephone surveys, and personal surveys. Each survey methodology has advantages and disadvantages. Keep in mind that self-administered survey results may not be generalizable. Mail, telephone, and personal surveys can all provide generalizable data if you have a valid and reliable list from which to select names, follow a statistical random sampling strategy, and have a high response rate.

For self-administered surveys, you might identify a group that meets regularly, go to that meeting, and ask group members to complete the questionnaire and return it to you. Such data collection occurs at club meetings, classes, and similar gatherings. Publications such as newsletters, magazines, and computer manuals include questionnaires for readers to fill out and return. Such self-administered surveys often suffer from low response rates, respondents who represent the extremes on survey topics, and data that lacks generalizability. Conversely, self-administered surveys are an inexpensive, convenient, and quick way of collecting data that can provide useful insights into specific issues as well as serve as an initial step in pretesting a questionnaire.

For mail surveys, you obtain a list of names, pull a sample from the list, send each person sampled a cover letter and questionnaire, and ask them to complete the survey and return it to you. Administering a mail questionnaire usually requires from three to eight weeks.

For telephone surveys, you obtain a list of names, pull a sample from the list, send an announcement letter to the people whose names were drawn, call them, and conduct the interview. Telephone surveys are usually more expensive than mail questionnaires, but you can collect more in-depth information with a telephone survey and can often generate a higher response rate if you follow appropriate procedures. While these surveys require a shorter time period, they do require more support staff and facilities. A major disadvantage of telephone surveys today is the telemarketing efforts that use "pseudo-surveys" to draw customers into their sales pitch. Therefore, it is imperative to immediately announce that your survey is not a marketing pitch and to demonstrate that early in the survey process.

For personal interview surveys, you obtain a list of names, pull a sample of names from the list, send an announcement letter to the people whose names were drawn, and then arrive in their homes or office to conduct the interview. Personal interview surveys are the most expensive of the survey techniques because they require field interviewers traveling to respondents' homes or places of business. To boost response rates in field surveys, the field interviewers usually return to the respondent's home or office three or more times. The one-on-one survey setting of personal interviews enables the interviewer to ask more in-depth questions, use needed probes, and observe the respondents. In most cases, one-on-one personal interviews are longer, provide more in-depth information, and increase the likelihood that respondents will not tire as easily as they do when answering telephone interviews. However, the high expense of personal interviews usually limits its use except in highly funded projects.

In recent years, survey researchers have been experimenting with using multiple strategies for gathering data. They may begin with a mail questionnaire, follow that with a telephone interview of nonrespondents, and then conduct personal interviews of those individuals who do not respond to either the mail survey or the telephone interview (Lavrakas 1993).

How Much Will My Survey Cost?

The cost of conducting a survey involves the interplay of fixed costs and variable costs. Fixed costs include those for the development of the questionnaire, pretesting the questionnaire, writing the codebook, and preparing the statistical analyses programs. Variable costs include those for the length of the survey instrument, the number of people to be surveyed, the nature of questions asked, the kind of data analyses required, and printing and production costs. Mail, telephone, and personal interviews incur additional expenses.

Mail surveys include the costs of envelopes for mailing the survey, postcards, and postage. The initial mailing packet includes a cover letter, a copy of the questionnaire, a return envelope with postage affixed, and the mailing envelope with sufficient postage for the weight of the package. Postage is best checked by preparing half a dozen mailing packages and then weighing them on electronic postal scales at the post office. You do not want your survey package to be delivered to the respondent with a postage due charge.

General cover letters should be addressed to each individual and printed on a quality printer to convey individual attention. For example, Dillman (1978) recommends that each cover letter be individually signed by the project administrator or researcher in blue ink to suggest that special attention is being paid to each respondent. Such details cost more, but they help boost the response rate.

For the return envelopes, use individually stamped and addressed envelopes rather than business reply envelopes. Jane Maestro-Scherer, director of Cornell University's survey research laboratory, says that using the stamped, addressed return envelopes helps boost response rates, as does printing the addresses on envelopes with a letter-quality printer to give the impression of individually typed letters.

Dillman's complete design method calls for at least two additional mailings beyond the initial mailing: 1) a postcard follow-up a week to 10 days after the initial survey, and 2) a second follow-up questionnaire packet with a new cover letter appealing to the respondent to complete the survey and return it as soon as possible. These additional costs must be calculated into the overall budget.

Telephone surveys include the costs of 1) telephone charges, 2) interviewer costs, and 3) announcement letter printing and postage. Local telephone surveys generate monthly charges for operating the telephone, and state or national telephone surveys also generate long distance charges. All telephone surveys usually require hiring several telephone interviewers whose costs include wages, training, and supervising. These interviewers may need to call potential respondents three or more times, at different times of the day, before making contact. Thus, each 20-minute telephone interview often requires about an hour of interviewers' time.

Survey researchers often use computer programs to generate telephone number lists from which they dial the sampled telephone number and conduct the survey (Lavrakas 1993). While these are commonly used on many topics, surveys relevant to scientific and technical issues often generate sampling from existing lists. When you sample from a list, send an announcement letter to the individuals sampled. Such letters help legitimize the survey, separate it from marketing ploys, and boost response rates. The costs of generating the mailing lists, as well as creating and mailing the announcement letter, add cost to the project.

Beyond basic production costs, the one-on-one personal interviews generate additional costs for the announcement letters and interviewers' costs. As with telephone surveys, sending an announcement letter to the individual selected helps boost response rates. Interviewer costs include training, hourly wages, mileage, meals, and lodging, if required. Depending on the location and length of the interviews, distance between interviews, and other logistical factors, a one-on-one personal interview can easily require three or more hours of an interviewer's time. The supervision of interviews, including the review of completed interviews, and occasionally, callbacks of the respondents, are additional costs of the personal survey.

Gaining Human Subjects Committee Approval

Under U.S. Government regulations, research involving human beings must be approved to ensure that the people subjected to study are not harmed. Thus, universities, colleges, government agencies, businesses, and research organizations have established human research committees. These committees usually consist of researchers, university or agency representatives, and community representatives, such as ministers. For each project that uses human beings as subjects, researchers must complete a human subjects committee form that provides details of the research, as well as information sheets and permission forms, and submit them for formal review. The committee then reviews the materials and makes recommendations to ensure that subjects understand the project and are willing to participate in it.

REFERENCES

Dillman, D. A. 1978. *Mail and telephone surveys*. New York: Wiley.

Fowler, Jr., F. J. 1990. *Standardized survey interviewing: Minimizing interviewer-related error*. Applied Social Research Methods Series, vol. 18. Newbury Park, CA: Sage.

Fowler, Jr., F. J. 1993. *Survey research methods*. Applied Social Research Methods Series, vol. 1. Newbury Park, CA: Sage.

Lavrakas, P. J. 1993. *Telephone survey methods*. 2d ed. Newbury Park, CA: Sage.

Long, M. D. Zimmerman, and G. Boirasky. 1993. A survey of the pollution prevention practices of small- and medium-sized Colorado businesses. Technical Report 94-1. Fort Collins, CO, Center for Research on Writing and Communication Technologies, Colorado State Univ.

Pollution Prevention Partnership. 1993. *Request for proposal to survey small- and medium-sized Colorado businesses on their pollution prevention practices*. Denver: Colorado Pollution Prevention Partnership.

CHAPTER

10

Identifying the Population and Sampling

When considering whom to survey, think about

- Identifying the population
- Assessing the quality of the list
- Selecting a sampling strategy
- Sampling

IDENTIFYING THE POPULATION

When you begin planning the sampling for a survey ask, "Whom will I be surveying? Why will I be surveying this group? Does a list exist from which I can pull a sample?" While many general polls and surveys seek to ascertain information about the general population, surveys also focus on specific groups. The problem itself often dictates the population that you will survey.

For example, the Pollution Prevention Partnership of Colorado was interested in helping small- and medium-sized businesses prevent pollution caused in such areas as 1) dry cleaning, 2) automotive and fleet maintenance, 3) printing, 4) metal fabrication, 5) microelectronics, 6) biomedical, and 7) aviation. More specifically, the Pollution Prevention Partnership was interested in surveying those individuals within the respective companies who were responsible for handling pollution prevention. After exploring several lists and querying the list publishers about how they developed the list, Long, Zimmerman, and Boirasky (1993) sampled names from the *Colorado Business Directory* because it contained the major businesses by industrial codes as well as the annual revenues of the companies and number of employees.

Once you have identified the population, consider whether or not a list exists from which you can sample. For general population surveys, survey researchers often rely on random digit dialing of telephone numbers generated by a computerized system. For a detailed discussion of using random telephone number generation, see Lavrakas (1993). For personal interviews, researchers and commercial survey companies have developed careful strategies for sampling communities. These strategies entail selecting specific blocks within a community and then sampling specific homes within those communities (Fowler 1993).

For specific populations, you will need specific lists. For example, Zimmerman, Muraski, and Peterson (1993) surveyed writers and editors who were members of the Society for Technical Communication for an in-depth study of technical communicators' roles. Accordingly, when your problem dictates specific groups, first look for professional organizations to which individuals of the specific population might belong. Such organizations usually have a newsletter, magazine, journal, or other regular publication that is distributed through a mailing list.

To find out whether a particular profession has a professional organization, check with individuals who work in the field. In addition, you can look for a professional organization, trade magazine, or newsletter in such publications as *Gale Directory of Periodicals*, *Standard Rate and Data*, and *Gale Encyclopedia of Associations* as well as national, regional, or local directories of special businesses, industries, government agencies, and organizations.

Once you verify that the professional organization has a mailing list, you will need permission to use it. Some organizations cooperate fully if the project will be in their organization's interests; others control access more tightly. Sometimes the president, executive director, or secretary will have authority to let you use the list. In other cases, the officer may need to request that permission from the executive committee, board of directors, or membership. Acquiring such permission may take a few days or more than a month.

Some organizations will allow you free use of their lists but will charge for generating the list. Others will charge a standard usage fee. Many organizations maintain their lists on databases that can be programmed to generate a random sample either as a computer file, printed labels, or both. Try to obtain the sample as an ASCII file. ASCII files can be imported into many computer programs where they can be used for generating cover letters, postcards, envelopes, and other correspondence.

ASSESSING THE QUALITY OF THE LIST

The quality of your final data can be no better than the quality of the list from which you sample. When selecting a list, look for one that is recent, up-to-date, identifies individuals by name, and provides addresses and other background information. Consider four factors when assessing list quality:

1. How was it developed?
2. Is the list representative of the target population?
3. How frequently is the list updated?
4. How does the organization maintain its list?

First, how was the list developed? Lists can be an organization's membership lists; individuals subscribing to a particular publication; individuals being licensed, such as medical doctors or nurses; or individuals who purchase products.

In some cases, you may need to generate a master list from which to pull a sample or develop an alternative sampling strategy. For example, the state of New York asked Cornell's survey research laboratory to conduct a survey of people who purchased New York fishing licenses. To generate the list, the laboratory staff obtained the books that contained carbon copies of fishing licenses. Then the laboratory staff sampled from the carbon copies, read and deciphered the names and addresses, and entered that information into a database. Developing such a list can be time consuming, difficult, and costly. In some instances, the source of the names may limit the quality of your list. Always carefully explore quality issues when considering lists.

Second, does the list represent everyone in a population or only a segment? For example, Zimmerman, Muraski, and Peterson (1993) surveyed members of the Society for Technical Communication (STC). While STC is the largest professional society of technical and scientific communicators, not everyone who is a technical communicator belongs to STC. In addition, while some 5,800 STC members indicated they were writers, editors, managers, or freelancers as of March 1990, the total membership actually numbered more than 12,000 in 1990. Thus, the survey results were generalized only to March 1990 STC members who indicated they were writers, editors, managers, or freelancers.

When you use a list, keep in mind that you must limit your findings to that list. At times, it may be impossible, if not economically unrealistic, to generate a master list of a target population and then sample from it. The cost of generating the master list may far exceed the budget of conducting the survey, and so you will, at times, find it necessary to sample from a less than ideal list. Work within your budget, identify and evaluate the list, and then sample from it.

Third, how frequently are the lists updated? Some organizations update mailing lists daily, weekly, monthly, or less frequently. To illustrate, an organization of communication educators appeared to have doubled its membership in four years, but the reality was that the list had not been updated for several years and, therefore, many nonpaying members' names had not been purged. Since that list included the names of individuals who were no longer teaching, who were no longer interested in the organization, who had died, or who may have changed professions, the value of that list would be questionable for a research project focusing on a topic of relevance only to current communication instructors. In contrast, should a researcher be interested in learning why some individuals had dropped out of the organization, the old list, even though it was "dirty," may have been useful.

Fourth, how does an organization maintain its list? Many organizations keep their lists on software programs such as word processing, spreadsheets, databases, or specialized programs, and a few may still maintain their lists with paper and pencil. Some software programs provide systems for drawing samples or can be programmed to draw such samples. A staff member who is proficient in using the database may be able to write a sampling program.

As the foregoing discussion indicates, a variety of issues may dictate the quality of the list. Thus, ask questions of the individuals who develop, maintain, and use the list to ascertain how they manage the list when you ascertain its quality.

SELECTING A SAMPLING STRATEGY

Once you have a list, you must decide whether to sample or conduct a complete census of the list. Conducting a census means surveying everyone on that list; sampling means selecting only a portion of the list to survey. If done properly, using a sample can save time and money, allow you to survey fewer individuals, and provide the opportunity to report a sampling error.

Sampling is divided into random and nonrandom sampling. A statistical random sample enables you to study a small number of people in the population, draw conclusions about them, and then generalize back to the larger population. For example, the systematic random sampling used to survey members of the Society for Technical Communication enabled the researchers to generalize back to all 5,800 writers, editors, managers, and freelancers in STC in March 1990 (Zimmerman, Muraski, and Peterson 1993). These findings had a sampling error of 3.5 percent. Thus, the researchers were 95 percent confident that the true responses of the population studied were within 3.5 percent points of the survey responses. For example, 61 percent of the individuals sampled were women, so the researchers could assert that between 57.5 percent and 64.5 percent of the population were women.

Nonrandom samples may be divided into purposeful samples and haphazard samples. For a purposeful sample, you select individuals who have selected characteristics. When you do this, you cannot generalize back to a larger population because you have no way of knowing if the individuals selected truly represent all individuals in the population. A college class represents a purposeful sample as does everyone attending a monthly week of a particular professional organization. When you use a purposeful sample, you have no way of knowing whether the individuals present at a particular meeting truly represent everyone in a group. Some of the absent individuals may be on field trips, sick, or out of town, any of which may indicate that they are quite different from individuals who attend a particular meeting. This is not to say that you should not use purposeful samples, but that you must realize the limitations of your data and not generalize the findings beyond those whom you surveyed.

Haphazard samples can generate precarious survey data. Consider the opening example in chapter 9 of the natural resource manager who sent students to interview lake users. If the natural resource manager was interested in extended camping practices of individuals visiting high mountain lakes, and the students interviewed day-users, fishermen, fisherwomen, and natural resource agency personnel who happened by, the data would be of questionable value.

SAMPLING

You can sample either manually or with computer assistance. Once you have pulled the sample, you should keep careful track of the names, addresses, and telephone numbers of the individuals sampled.

Manual Sampling

Assume you have the latest directory of XYZ association with 15,000 names, arranged by alphabetical order of last name, and published last month. Assume further that your budget will allow you to sample 500 names. You could use a table of random numbers to randomly select 500 individuals and then search the list for the names selected. A more expedient way, though, is to systematically random sample from the list of 15,000. To systematically random sample, you start with a random number and draw every nth name. First, establish the sampling interval; then, select the random number within the interval; and then, starting with the random number, sample every nth individual on the list. To illustrate, assume you want to draw a systematic random sample from the XYZ directory. To find the interval, divide the 15,000 by the 500. You now have an interval of 30. Next, you go to a table of random numbers in a standard statistics textbook and select a random number between 1 and 30 following the random number procedure specified in the statistics book you are using. Many of the standard research methods and statistics textbooks, such as Babbie (1992), Nachmias and Nachmias (1987), and Vernoy and Vernoy (1992), have tables of random numbers.

If you have a statistical calculator, such as the Texas Instruments TI-35 Plus, or a business calculator, such as the Texas Instruments BAII Plus, you can automatically generate random numbers. On both units, you turn on the machine, clear it, press "2nd" and then "Rand." The calculator displays the random number. Some calculators generate random numbers between 0 and 1 and express the numbers as a decimal, such as 0.61. You can convert 0.61 to 61 for sampling procedures. In the example, you would repeat the process until you obtain a random number between 1 and 30. Statistical computer programs, such as spreadsheets and database programs, also have random number generating functions. If you have such software, review the manuals and the online help functions.

Assume you followed one of these procedures and obtained a random number of 28. To draw the random sample, you would select number 28 as the beginning point and then you would count every 30th name in our sample—28, 58, 88, and so forth—until you have drawn your sample of 500 names. Once you have the sample, you should enter it into a word processing file program, a database, or a spreadsheet to help you track the individuals while you administer the survey. Be sure to enter the information correctly, because you will use it to address letters, envelopes, and postcards when collecting your data.

REFERENCES

Babbie, E. R. 1992. *The Practice of Social Research*. 6th ed. Belmont, CA: Wadsworth.

Fowler, Jr., F. J. 1993. *Survey research methods*. 2d ed. Applied Social Research Methods Series, vol. 1. Newbury Park: CA: Sage.

Lavarkas, P. J. 1993. *Telephone survey methods*. 2d ed. Newbury Park, CA: Sage.

Long, M., D. Zimmerman, and G. Boirasky. 1993. A survey of the pollution prevention practices of small- and medium-sized Colorado business. Technical Report 94-1. Fort Collins, CO, Center for Research on Writing and Communication Technologies, Colorado State Univ.

Nachmias, D., and C. Nachmias. 1987. *Research methods in the social sciences*. New York: St. Martin's Press.

Standard Rate and Data Service. 1993. *Consumer magazine and farm publication rates and data*. Skokie, IL: Standard Rate and Data Service.

Troshynski-Thomas, K., and D. M. Burek, eds. 1994. *Gale directory of publications and broadcast media*. Detroit: Gale Research.

Vernoy, M.W., and J.A. Vernoy. 1992. *Behavioral statistics in action*. Belmont, CA: Wadsworth.

Zimmerman, D. E., M. Muraski, and J. Peterson. 1993. Who are we? A look at the technical communicator's role. Technical report. Fort Collins, CO, Center for Research on Writing and Communication Technologies, Colorado State Univ.

CHAPTER

11

Developing Questionnaires

An effective survey presents understandable questions to the respondent. Someone inexperienced in developing and handling surveys can easily go astray in developing and ordering questions for the survey instrument.

Developing the questionnaire requires a careful approach that includes

- Drafting questions
- Avoiding faulty question construction
- Ordering questions
- Pretesting the questionnaire

DRAFTING QUESTIONS

Chapter 9 stressed the importance of developing detailed objectives to guide your questionnaire development. The more detailed you can make your objectives, the easier it will be to develop your questions.

To learn how to draft questions, consider the differences between open-ended and closed-ended questions. Open-ended questions ask the interviewee to provide the response; closed-ended questions provide limited responses.

Assume you were developing a survey to learn more about continuing education and the professional development of technical communicators. You might consider the following format.

> Some technical communicators complete additional training while others do not. Have you completed any professional development training in the last two years?

____ No ____ Yes. If yes, please describe the kinds of professional development training that you have completed:

____

____

The yes/no portion of the question is a closed-ended question while the second part is an open-ended question. When you ask open-ended questions, you must be sure that your audience is articulate enough to provide detailed responses. Keep in mind too that open-ended questions are more difficult to code and analyze than close-ended questions. (This issue is covered in chapter 13.)

Closed-ended questions can take many different forms. Responses may include different, fixed categories. To illustrate, consider revising your question to the following presentation.

For each of the professional development activities listed below, check those that you took part in during the last year.

____ Monthly professional meetings

____ Regional professional meetings

____ National professional meetings

____ One-day workshops or seminars

____ Two- or three-day workshops or seminars

The following type of questioning uses a scale response to measure the kinds of comments that faculty members make on students' writing.

When you assess students' written assignments, how frequently do you comment on the following topics?

	Hardly Ever				Most of the time
a. Content	1	2	3	4	5
b. Organization	1	2	3	4	5
c. Clarity	1	2	3	4	5
d. Conciseness	1	2	3	4	5
e. Spelling	1	2	3	4	5
f. Grammar	1	2	3	4	5
g. Style	1	2	3	4	5
h. Slanting for readers	1	2	3	4	5

Or, consider this scale question with five response categories.

Some people tell us that computers have made writing harder for them, while other people tell us computers have made writing easier for them. Would you say computers have made writing

____ A lot easier for you
____ Somewhat easier for you
____ Neither easier nor harder for you
____ Somewhat harder for you
____ A great deal harder for you

Generally, the multiple response questions provide more insight into activities, behaviors, or attitudes because they allow for a wide range, or a variance, in response. Variance in responses tells you more about a concept than dichotomous response categories such as yes/no questions.

To use closed-ended questions, you need to understand enough of the issues surrounding the question to develop response categories that the interviewees can understand. Keep in mind too that self-administered questionnaires and personal interview questionnaires can use more response categories than telephone surveys, where the interviewee will have trouble remembering multiple categories.

AVOIDING FAULTY QUESTION CONSTRUCTION

It is easy to run astray when asking questions if you use

- Pseudo-data questions
- Leading questions
- Loaded questions
- Double-barreled questions
- Subject-specific jargon
- Vague terminology
- Complex questions

The following discussion illustrates common problems associated with question construction.

Pseudo-Data Questions

Stanford professor and communication researcher Steve Chaffee advises against asking questions on topics that respondents have not previously considered (Chaffee 1971). Too often when people develop surveys on specific topics, they do not ask whether the respondents have thought about a topic.

The respondents, for their part, may often provide socially desirable responses to questions they had not previously considered.

For example, suppose you were surveying a general population about their concern for aardvarks. Many people may not know what an aardvark is, and even fewer would have thought about them in recent weeks. Too often, when too close to a subject, some researchers fail to realize that they are asking people to answer aardvark questions—questions on topics that they either know very little about or have thought little about. When queried, however, many people try to help and answer researchers' questions, thus creating what Chaffee coined "pseudo" data.

In addition to being concerned about pseudo-data, you also need to consider what impression your questions will produce. For example, a citizen received a telephone call from a pollster trying to ascertain his viewpoints on a candidate running for reelection in the district. This citizen had little knowledge of the legislator. After being asked a series of questions, the citizen told the pollster, "I really didn't know much about representative X until you called, but as a result of your questions, I sure won't vote for him."

A strategy for overcoming questions that create problems for your respondents is to give social license, or permission, to acknowledge that they have not thought about the question or are not interested in the topic. For a survey about aardvarks, you might say:

> Some people have thought about aardvarks in recent weeks, while other people have not thought about aardvarks in recent weeks. Have you thought about aardvarks in recent weeks?
> ___ Yes (Go to question 2) ___ No (Go to question 6)

By directing respondents to question 6, you avoid forcing them to answer questions 2 through 5, which address specifics about aardvarks.

Leading Questions

Leading questions are those that are structured so that they signal the desirable answer to the person being interviewed. A typical leading question might be:

> You think aardvarks should be preserved, don't you? ____ Yes ____ No

While some people might be willing to say "no" to the above aardvark question, other topics lend themselves to high levels of agreement. Consider a question on reducing taxes.

> You think that we should reduce personal income taxes, don't you?
> ____ Yes ____ No

What person would favor increasing their taxes? The question by its very nature leads respondents into predetermined responses.

Loaded Questions

Loaded questions force responses that might not be true. For example,

Have you stopped beating your spouse?

An answer of "no" suggests that the respondent continues to abuse the spouse, while an answer of "yes" says that the individual used to, but no longer does so. With loaded questions, the respondent may not fully understand the question, and thus answers with an understanding of part of the question.

Other examples of loaded questions might be

Would you favor making computer programs easier to use?
____ Yes ____ No

Would you support keeping guns out of the hands of criminals?
____ Yes ____ No

Would you support better regulations to protect the environment?
____ Yes ____ No

These last two questions are both leading and loaded questions because they suggest what appears to be socially acceptable and desirable behaviors to which many respondents would be hesitant to answer "no." Only strongly committed respondents might answer "no" to the above questions.

Double-barreled Questions

Double-barreled questions involve asking two questions at the same time. In a survey of personal computer users, you might ask

Do you use word processing and spreadsheets on your computer?
____ Yes ____ No

How would the respondent who uses *only* word processing programs respond? How would the respondent who uses *only* spreadsheets respond? If a respondent answers "yes," how will the researcher know whether the response represents both or only one or the other? To avoid double-barreled questions, develop individual questions for each category that you want to explore.

Questions with Subject-Specific Jargon

For some topics, respondents may not be familiar with selected technical, scientific, or specialized terms. For example, the Poudre R-1 school district, Larimer County, Colorado, surveyed residents about developing a strategic plan (Hendrix 1991). The survey asked respondents to rank the importance of each question on a 1 to 5 scale with 1 being "not at all important" and 5 being "very important." The opening question asked respondents to assess the importance of the following:

> To develop *exit outcomes for graduates* including periodic benchmark indicators, supporting assessment and reporting systems. (Common standards for graduation.)

What does *exit outcomes for graduates* mean? If respondents are unfamiliar with educational jargon, such questions may leave them bewildered and confused. What are periodic benchmark indicators? Supporting assessment and reporting systems? To really answer such questions requires an understanding of the educational jargon or having that jargon translated into understandable terms.

Vague Terminology

Vague general terms may have different meaning to different people. For example, the Poudre R-1 survey also asked respondents to rank the importance of the following items:

> To ensure high levels of language, literacy, and information and communication skills using a variety of media.
> To ensure high levels of problem solving skills and symbolic logic in mathematics, science, and technology.

On both items, how does the school district define "high level"? What is a high level to one person may well be a low level to another individual. These are relative terms that cannot be precisely defined and so should be avoided.

Complex Questions

Most people find long, complex questions hard to understand. Generally, although not always, longer sentences are more difficult to understand than shorter sentences. And the more different ideas that are presented within a question, the longer the mental processing time required to understand the question.

Thus, questions that force respondents to consider a complex scenario may leave them either simply confused or at risk of misinterpreting the question. Suppose you were interested in learning more about the ways in which people control their weight, and you wanted to know what circumstances lead people to diet. So you devised a draft question such as

> If you were overweight and interested in losing more than 15% of your body weight, would you consider a combination of approaches that included a prolonged dieting sequence of reducing your caloric intake by more than 20% daily,

while increasing your exercise efforts to burn 30% more calories than you currently do?

Such an opening sequence provides too many different propositions for people to understand and consider.

Consider question six from Ross Perot's March 1993 United We Stand ballot (page 1993).

6. If our government wants the American people to pay more taxes, should it provide leadership by example—all sacrifices begin at the top—by cutting Congress' and the President's salaries by ten percent and reducing their retirement plans to bring them in line with those of the American people?
____ Yes ____ No

Not only has Perot provided a complex question, but also a leading and loaded question. Few people would not favor seeing elected officials lower their salaries and cut their retirement plans.

ORDERING QUESTIONS

As you develop a questionnaire, consider

- Engaging the interviewee
- Placing potentially sensitive questions late in the survey
- Avoiding sensitizing interviewees to subsequent questions
- Moving smoothly from question to question

Engaging the Interviewee

The introduction and opening questions should interest the respondents in the survey, not pose any threatening questions, and set the stage for the subsequent questions. Begin by asking questions on topics relevant to respondents and ones in which they are knowledgeable.

Placing Potentially Sensitive Questions

Avoid placing potentially sensitive questions early in the survey. Over the years, survey researchers have found that demographic characteristics such as age, education, income, and related characteristics can be sensitive questions.

Professor Bud Sharp, former director of the University of Wisconsin Survey Research Laboratory, tells of a survey of women 18 to 44 years of age. While respondents provided detailed answers about a wide range of issues about intercourse, the use of contraceptives, and sexual practices, several respondents balked at answering questions about education and income, explaining that that information was "too personal."

Experienced survey researchers place questions gathering the key or important information early in the questionnaire and place potentially socially sensitive and demographic questions toward the end of the questionnaire. Thus, should the respondents refuse to answer the sensitive questions and terminate the survey, the interviewer will already have gathered the key information.

Avoiding Sensitizing Interviewees to Subsequent Questions

When ordering the questions, consider whether asking some questions will sensitize respondents to subsequent questions, or give them clues to the desired answers for questions later in the survey. Such an approach means planning the sequence of questions and then stepping back to ask, "Will asking question X clue the respondents into a desired answer later in the survey?"

To illustrate, reconsider the national survey of technical communicators (Zimmerman, Muraski, and Peterson 1993). The researchers wanted to know how technical communicators saw their changing roles and how they understood their current roles and positions. The researchers reasoned that the questions about the future role of technical communicators should come before the in-depth questions that concentrated on current role characteristics (panel 11.1). Note that they first asked two screening questions and that they used open-ended questions to avoid forcing responses into previously defined categories.

Moving Smoothly from Question to Question

With the general order of the questions devised, concentrate on making smooth transitions from one area of the questionnaire to the next. Consider using subheads that clearly divide and label sections as well as provide transitions such as

> Now we'd like to ask about your current position . . .
>
> Now we'd like you to tell us about the information sources that you regularly use for your job.

Such transitions signal the changes in questions for the respondent and provide a smooth flow to the next section.

PRETESTING THE QUESTIONNAIRE

Once you have a working draft of the questionnaire, carefully review it for possible errors, make the needed corrections, and then make copies for pretesting. First, have three to six people familiar with the topic complete the

PANEL 11.1

THE OPENING PAGE OF A NATIONAL SURVEY OF TECHNICAL COMMUNICATORS' ROLES (ZIMMERMAN, MURASKI, AND PETERSON 1993)

1. Some technical communicators have thought about how their roles are changing. How frequently have you thought about how technical communicators' roles are changing?
   ___ Hardly ever ___ Seldom ___ Sometimes ___ Often ___ Very often
2. How much do you expect technical communicators' roles to change in the coming decade?
   ___ No change ___ Hardly at all ___ A little
   ___ Some ___ A great deal ___ Don't know
3. What kinds of new knowledge do you think technical communicators will need to acquire to perform their jobs in the coming decade? Please describe/ explain.
   ____
   ____
   ____
4. What kinds of new skills do you think technical communicators will need to acquire to handle their jobs in the coming decade? Please describe them.
   ____
   ____
   ____
5. What other changes do you envision for technical communicators in the coming decade? Please describe them.
   ____
   ____
   ____
6. What do you think will be the best ways for technical communicators to prepare for these changes? Please explain.
   ____
   ____
   ____

questionnaire, and keep track of the time it takes them to do so. If it's a telephone or personal survey, then conduct the interview as it will be done. Once they have completed the pretest, debrief them. Ask questions such as

- Did you find the questionnaire interesting?
- Did the questions flow smoothly from one section to the other?
- Were you familiar with the topics covered?

- Did you have any trouble understanding specific questions?
- Did you find any unfamiliar terms?
- Would you use different words to phrase any of the questions? If so, which ones? Why?
- What was your reaction to the questionnaire's length?
- What do you think about the order of the questions?
- Did any of the questions clue you into, or suggest answers to, subsequent questions?
- Did you find any questions that you thought "should" be answered in a particular way? Did you give what you thought were the socially correct answers?
- Do you think people in the target population will have any problems with the questionnaire? With individual questions?
- Did you find any of the questions especially sensitive or too personal?
- Do you have any other suggestions for improving the questionnaire?

After you have finished interviewing the pretest respondents, look carefully at their answers. Tally the pretest respondents' answers to each question. Next, ask yourself the following.

- Is there dispersion in the answers?
- Do any questions fail to produce variance in the responses?
- Are the answers to the open-ended questions as expected?
- Do the responses to closed-ended questions align with expectations?
- As presented, will the questions provide the kind of data needed from the survey? If not, what changes may need to be made?

If the needed changes are minor, based on your assessment of the initial pretest, revise the questionnaire and then conduct a second pretest using a larger pretest group of respondents. Some researchers either pull a small sample of 20 to 30 subjects from the population to be surveyed, or purposely select 20 to 30 subjects with widely ranging differences on key characteristics.

For example, Zimmerman, Muraski, and Peterson (1993) conducted three pretests for the national survey of technical communicators. They first selected half a dozen local technical communicators to complete the survey and provide feedback. Next they asked a dozen graduate students in a technical communication M.S. degree program to complete and critique the survey. Finally, they sent the survey across the country to technical communicators they knew who had 2 to 25+ years of technical communication experience. Based on these pretests, the researchers shortened the questionnaire, added some questions, and revised others.

Some researchers run a complete data analysis on the responses and look carefully for the dispersion of responses. By carefully working the pretest questionnaire through coding and data analysis, you will find it easier to handle the final survey data.

After conducting the pretests and analyses, rewrite and revise the questionnaire as needed. You may find it necessary to reorder some items, shorten or expand other items, change terminology, or, in other ways, recast the instrument based on what you learned during the pretests.

REFERENCES

Chaffee, S. H. 1971. Pseudo-data in communication research. Symposium on Coorientation. Communication Theory and Methodology Division. Association for Education in Journalism. Columbia, SC. August 1971.

Hendrix, D. 1991. *Community Survey*. Fort Collins, CO: Poudre R-1 School District.

Page, S. 1993. Now, here's the thing. *Fort Collins Coloradoan*, Sunday, March 21, 1993. Insight section, E, p. 1.

Zimmerman, D. E., M. Muraski, and J. Peterson. 1993. Who are we? A look at the technical communicator's role. Technical report. Fort Collins, CO, Center for Research on Writing and Communication Technologies, Colorado State Univ.

CHAPTER

12

Administering the Survey

After developing the questionnaire, you will need a strategy for announcing the survey to the individuals whom you hope to survey. The announcement will help to legitimize your research project and to enlist the help of those whom you approach. Doing so also helps overcome the possible perception that your survey efforts may be part of a marketing strategy.

The following discussion covers administration of group administered surveys, mail surveys, and telephone and one-on-one personal surveys.

HANDLING GROUP ADMINISTERED SURVEYS

Advance notification takes different forms.

For self-administered surveys of classes, request permission of the class instructor several weeks prior to administering the survey in class. Keep in mind that since a survey could easily take 10 to 20 minutes or more of the class time, instructors may need to give up instructional time. In approaching instructors, you need to explain the project and the value, if any, the project might be to class members. When you make arrangements to administer the survey, double-check the time, date, and room location. If you have made these arrangements several weeks in advance, call a couple of days before the agreed date to verify the time.

If you will be administering the survey to a group, club, or other gathering, you will need to seek permission several weeks in advance. In some cases the president of the organization can grant you permission; in other cases the executive board or key officers will provide the approval. Such approval may take a few days or more than a month depending upon the organization, the regularity of its executive board meetings, and other activities.

Once you secure permission, the process works similarly for both classes and groups. Before you arrive, prepare a succinct presentation that 1) explains the project in general terms, 2) tells how the individuals were selected, 3) tells why their participation is important, 4) assures confidentiality of their responses, and 5) appeals to them to participate by completing the survey. When you prepare your narrative, keep it short. You should not provide a detailed explanation of the project because you want to avoid undue sensitization and you do not want to bias responses. Panel 12.1 provides a sample letter that may help you in preparing your presentation.

If your survey may result in publication or requires approval of your school's human research committee, you will need to explain the information sheet and the participant sign-off sheet. While the specifics of these vary from school to school, they become a necessary part for some projects.

To conduct a classroom survey, you will need to prepare a packet that includes the information sheet, the subject participation agreement, the cover sheet for the questionnaire, and the questionnaire. The cover sheet for the questionnaire should request information about the participant and have a place for inserting a control number (panel 12.2). The first page of the questionnaire should have a blank for the control number and include the questions that introduce the project (panel 12.3).

Once participants complete the permission forms, cover sheet, and questionnaire, you will need to process the packet. Remove the permission forms and store them separately from the questionnaires. Add a control number to the cover sheet and place the same control number on the first page of each questionnaire. Then remove the cover sheet and store it separately from the questionnaires. To maximize the confidentiality of responses, do not put any name, social security, student number, or other information beyond the control number on the questionnaire.

HANDLING MAIL SURVEYS

A good cover letter plays the key role in gaining maximum participation in mail surveys. Dillman (1978) developed a method for drafting cover letters and executing mail surveys, called the Total Design Method, that produces response rates of 60 to 90 percent. According to Dillman, the ideal cover letter is a one-page letter typed on official letterhead that provides the date mailed; that is addressed to a specific individual; that tells what the study is about and its social or professional usefulness; that explains why it is important for the recipient to participate; that promises confidentiality; that offers a token reward, such as a summary of the results; that provides directions should questions arise; that offers words of appreciation; that provides a signature block with the project director name and title; and that is signed in blue ink by the project director. Panel 12.4 provides an example of the cover letter used for the Society for Technical Communication national survey

PANEL 12.1

COVER LETTER OF THE ELECTRICAL ENGINEERING STUDENT SURVEY

October 15, 1993

Electrical Engineering Students

Department of Technical
Journalism
Fort Collins, Colorado 80523
(303) 491-6310 or 6319

Dear Electrical Engineering Major:

We're seeking your help to enhance the teaching of professional communication at Colorado State University.

Specifically, we're conducting a study that looks at your prior writing and computer experiences, and what you think about writing and computers.

We would like you to volunteer to help us better understand how you write and use computers by completing a survey outside of class and returning it during the next class period. Your volunteering or refusing to take part in this study will not influence your grade in this course.

First, we'd like you to complete the attached address and telephone number card as well as two copies of the consent form. We need your address and telephone number so that we might call to remind you to return the questionnaire should you forget to return it next class period. Keep a copy of the consent forms for your records.

Your responses will remain confidential.

If you're willing to take part in this study, please remove this cover letter, complete the attached consent form and the address information sheet, and return them to us.

Thank you for helping with this project.

Sincerely,

Donald E. Zimmerman
Professor

PANEL 12.2

COVER SHEET OF THE ELECTRICAL ENGINEERING STUDENT SURVEY

Cover Sheet **Writing Study—Part II**

Control Number 932-_____-_____-_____

Name ____

ID Number ____

Class ____

Class Meeting Times ________________________________

Date ____

This project is designed to learn more about you as a developing writer and communicator, how you use computers, and how we might enhance communication instruction.

Please answer the following questions openly and as completely as you can. Your answers will help us teach communication more effectively. Your responses will not influence your grade.

We need your student identification number in case we need to ask you additional questions about your developing communication and computer use skills later.

Your answers will remain confidential and your name will not be used in any report or article that we might publish from the data being collected.

(Zimmerman, Muraski, and Peterson 1993). Jane Maestro-Scherer, director of Cornell's survey research lab, recommends using a daisy-wheel printer or a laser printer to address envelopes so that each envelope looks individually typed. Dot-matrix printers or labels imply bulk-mailing.

To conduct the mail survey, send each participant a package consisting of the cover letter, the questionnaire, and a stamped, addressed, return envelope. Make sure that you have sufficient postage on the package envelope so that the package does not arrive with postage due in the respondent's mailbox. Each questionnaire should carry an identification code for tracking purposes, as illustrated in panels 12.3 and 12.5. This initial mailing may produce up to a 30 percent response rate.

Dillman recommends that a postcard be mailed to everyone 7 to 10 days after the initial mailing. The postcard should provide the date mailed, create a link with the earlier letter and questionnaire packet, offer gratitude to those early respondents, reiterate why the survey is important, provide an invita-

PANEL 12.3

FIRST PAGE OF QUESTIONNAIRE SHOWING CONTROL NUMBER

(A survey instrument for personal interviews of faculty members about writing assignments in their respective classes.)

Part I. Assignments **Control Number 933- ____ -____ - ____**

1. What courses do you teach?

2. What assignments do you give students in each course?

3. Are any of them communication assignments?
   3a. Written assignments?
   (LIST ASSIGNMENTS)
   1.
   2.
   3.
   4.
   5.

   3b. Are any assignments collaborative or team writing? Yes No

   3c. Do you provide students written guidelines for these assignments?
   Yes No
   If yes, would you please provide us with a copy?

   3d. What specific guidelines do you give students to direct their work on the assignments?

tion to get a replacement questionnaire, and, again, show the project director's individually penned signature in blue ballpoint ink below the signature block (panel 12.6). The postcard may generate an additional 10 to 25 percent percent response rate.

A third mailing, from 14 to 21 days after the initial mailing, will encourage even more respondents to complete the questionnaire. The third mailing should consist of a new cover letter, questionnaire, and stamped and addressed return envelope. The cover letter in the third mailing should include the date, link the packet to the previous communications, explain the usefulness of the study, reconfirm the importance of the recipient to the study, and provide a signature block signed in blue ink by the project director (Dillman). Panel 12.7 provides an example of the third mailing. Send the third mailing only to those individuals who have not responded to the earlier mailings.

Careful tracking of the respondents is key to the success of the repeated mailings. You do not want to send another questionnaire to someone who has returned the questionnaire. This is where the control number of the ques-

PANEL 12.4

COVER LETTER FOR THE STC SURVEY

April 30, 1990

Name
Address

Department of Technical
Journalism
Fort Collins, Colorado 80523
(303) 491-6310 or 6319

Dear ___ :

Technical communicators' jobs and roles have changed greatly in the last decade, and more changes may emerge in the 1990s.

To enhance the understanding of technical communicators' roles, professionalism, and information sharing, we are conducting a national study of technical communicators.

Recognizing the importance of this study, the Society for Technical Communication has provided a research grant to Colorado State University's Department of Technical Journalism for the project.

You are one of 500 randomly selected members of the Society for Technical Communication whom we are asking to help. So that the results will be truly representative of technical communicators in STC, please complete the enclosed questionnaire and return it promptly to us in the enclosed stamped, pre-addressed return envelope.

Your responses will be confidential. We have printed an identification number on the questionnaire so we can check your name off the master mailing list. Your name will never be placed on the questionnaire or mentioned in any published results.

We will provide a summary of the study results to STC's board of directors, and we will write a series of articles for STC's journal, Technical Communication. To receive a summary of the study's results print your name and address on the back of the return envelope.

If you have questions, feel free to write or call (303-491-5674). Thanks for your help.

Sincerely,

Don Zimmerman & Jodi Wolf
Project Directors

PANEL 12.5

First Page of the STC Mail Questionnaire

Exploring Technical Communicators' Roles in the 1990s

A National In-depth Survey of Technical Communicators

Conducted by
Communication Research Laboratory
Department of Technical Journalism
Colorado State University
Fort Collins, CO 80523

Supported by a Research Grant from
The Society for
Technical Communication.

Control No. _____

tionnaire provides an effective management tool. Print a master list of each person who was sampled, including their name and address, and add the control number to each name on the master list. When a questionnaire comes in, you can mark the respondent's name off of the list. When you are ready for

PANEL 12.6

POSTCARD MAILER FOR THE STC SURVEY

May 21, 1990

About a week ago we mailed you a questionnaire seeking your opinion about the roles of technical communicators.

If you have returned the questionnaire to us, please accept our sincere thanks. If not, please return the questionnaire today. Because we sent the questionnaire to a small sample of STC members, your answers are important so that the results are representative of STC members.

If you did not receive the questionnaire, or it was misplaced, please call us collect (303-491-5674) right away, and we will send another questionnaire today.

Thanks for your help.

Sincerely,

Don Zimmerman & Jodi Wolf
Project Directors

the subsequent mailings (postcards and second questionnaire packet), screen the list and delete the names of individuals who have completed the questionnaires before doing a second mailing.

The advancements of word processing and database software for personal computers can greatly facilitate the handling of surveys and the management of files for subsequent mailings. See the respective manuals for the different software to learn how to track mailings and handle the mail merges for addressing envelopes and letters.

Advances in electronic mail systems and fax systems can speed the delivery of your letters. For example, in conducting an evaluation of its editorial services, the Office of Information Transfer, U.S. Fish and Wildlife Service, sent e-mail messages to researchers encouraging them to complete and return the questionnaire.

As technologies continue to advance, new and innovative ways will be developed to distribute surveys and cover letters and to encourage responses.

HANDLING TELEPHONE SURVEYS

With today's increasingly heavy use of telemarketing, you will immediately need to dispel the idea that your survey is a marketing sales pitch. To begin, you should send an announcement letter to each person sampled to legitimize the survey and increase response rates. Lavrakas (1993) reports that advance notification letters are especially useful for elite populations.

PANEL 12.7

THIRD MAILING LETTER OF THE STC SURVEY

May 10, 1990

Address

Colorado State University

Department of Technical Journalism
Fort Collins, Colorado 80523
(303) 491-6310 or 6319

Dear ————:

About three weeks ago we wrote you seeking your help in a national study of technical communicators. As we noted, technical communicators' jobs and roles have changed in the last decade, and even more changes may emerge in the 1990s. Understanding where we are and where we are going as a profession can help us prepare for the coming decade.

Recognizing the value of this study, the Society for Technical Communication has provided a research grant to Colorado State University's Department of Technical Journalism. The research project is being handled by the Communications Research Laboratory in the Department of Technical Journalism.

As of today we have not received your completed questionnaire.

We are writing you again because of the significance each questionnaire has to the usefulness of the study. Your name was drawn through a scientific sampling process in which each technical communicator in the Society of Technical Communication's list had an equal chance of being selected. This means that only one in every 20 STC members is being asked to complete this questionnaire.

So that the results are truly representative of technical communicators, it is essential that each STC member in our sample return the questionnaire. As mentioned in our last letter, it is important that you complete the questionnaire.

In the event that your questionnaire has been misplaced, a replacement has been enclosed.

Your help is greatly appreciated.

Cordially,

Don Zimmerman & Jodi Wolf
Project Directors

Like the mail questionnaire cover letter, the announcement letter should follow Dillman's recommendations, outlined in the previous section. In addition, the letter should explain when a telephone interviewer will call to conduct the survey or to set up a convenient time for the telephone survey. The letter should be mailed so that it arrives two to five days before the respondent receives the initial telephone call. In that way, the individual will be more likely to recall the letter and respond positively to the interviewer's inquiries.

The opening minutes of your telephone survey should 1) identify the interviewer, the interviewer's affiliation, and the survey's sponsor; 2) provide a link with the announcement letter; 3) provide a brief explanation of the survey and sampling scheme; and 4) encourage participation.

Careful selection of telephone interviewers, whether volunteers or paid, plays a vital role in generating useful and helpful data. All interviewers must use a standardized approach, using the same introduction and asking the same questions as written on the questionnaire. Should the respondents ask questions, or the interviewers need to probe to ascertain additional information, all interviewers must respond in the same way.

Lavrakas (1993) provides a detailed discussion of training interviewers that ensures standardized consistency. Training should include the background of the survey, how to encourage respondents to participate, how to maximize participation, how to handle probes, and how to handle difficult respondents. As part of the training, interviewers should conduct several practice interviews before conducting the actual telephone surveys.

During the telephone surveys, a supervisor must monitor the interviewers, review completed questionnaires, and handle any needed callbacks to obtain missing or confusing responses.

HANDLING ONE-ON-ONE PERSONAL INTERVIEWS

The coordination of one-on-one field interviews is even more complex than handling telephone interviews. The time required to collect data quickly mounts for personal one-on-one interviews because the interviewers must travel to multiple locations to complete their assigned interviews and, in some cases, they must return several times if the individual is not at home or the office.

In addition to the components required for the telephone interview letter, the advance letter for field interviews should tell the respondent when to expect the interviewer and how interviewers will identify themselves. The announcement letter should arrive two to five days before the interviewer arrives.

When the interviewer arrives on the respondent's door-step, they should have with them a copy of the letter and some identification. Their dress and approach should be so that they appear unthreatening and their manner should be pleasant and easygoing. As with telephone interviewing, the careful selection of field interviewers, whether volunteers or paid, plays a vital role in generating useful and helpful data.

The administration of a crew of field interviewers is even more complex than the administration of telephone interviewers. Not only must the individual coordinating the field interviewers be concerned with the logistics of conducting the interview, but the coordinator may need to make arrangements for meals, lodging, and transportation, depending upon the nature of the survey.

The project coordinator must carefully review the completed interviews and monitor the interviewing progress. Lavrakas provides details on interviewer selection, training, and monitoring, as well as coordinating field interviewers.

EMERGING APPROACHES TO SURVEY RESEARCH

As with any scientific or technical discipline, researchers continue to develop new techniques and methods, reassess old techniques, and compare different methods. Chief among survey research changes are advances in sampling procedures, comparisons of different survey techniques, and data analysis advances.

For surveys of the general population, researchers have developed and evaluated a variety of alternative sample procedures. Lavarkas provides an extended discussion of the Kish, birthday, and Trodhalh-Carter-Bryant (T-C-B) sampling methods and other systematic approaches. Because the effectiveness of all methods is open to debate, researchers continue to assess the validity and reliability of different methodologies. Whatever the method used, the key issue is whether or not you can generalize back to the larger population and the accuracy of your findings.

In recent years, researchers have been exploring the use of combining survey methodologies to collect data. Some researchers have used an initial mail survey methodology, followed by a telephone interview of nonrespondents to the mail survey, and then a one-on-one personal interview of those individuals refusing to participate in either the mail or the telephone surveys. Such extended efforts can be time consuming and costly, but do generate a high response rate. Carefully weigh the costs of such undertakings against the potential benefits.

While computers have long been used for data analysis, techniques for using the personal computer and software for survey research is growing.

Researchers are developing computer aided telephone interviewing (CATI) software to speed the processing of information, and with the advent of laptop computers, researchers are now having interviewers use similar programs for one-on-one personal interviews in the field. These systems enable researchers to bypass some steps in the data analysis.

As computers become more and more powerful, advanced data analysis can now be completed on a personal computer with statistical analysis software. In other cases, spreadsheets and database programs can provide the needed data analyses. Because of the complexities of data analysis, the next chapter is devoted to data analysis techniques, including an extended example of one data set.

REFERENCES

Dillman, D. A. 1978. *Mail and telephone surveys*. New York: Wiley.

Fowler, F. J. 1993. *Survey research methods*. 2d ed. Applied Social Science Research Methods Series, vol. 1. Newbury Park, CA: Sage.

Lavrakas, P. J. 1993. *Telephone survey methods: Sampling, selection and supervision*. 2d ed. Applied Social Science Research Methods Series, vol. 7. Newbury Park, CA: Sage.

Zimmerman, D. E., M. Muraski, and J. Peterson. 1993. Who are we? A look at the technical communicator's role. Technical report. Fort Collins, CO, Center for Research on Writing and Communication Technologies, Colorado State Univ.

CHAPTER

13

Analyzing and Interpreting Data

Once the surveys are completed, making sense of the data requires a systematic and methodological approach that enables you to summarize and interpret the findings. This approach is a multistep process that includes

- Considering data analysis techniques
- Developing the codebook
- Entering the data
- Writing the software program
- Running the program
- Presenting your findings

CONSIDERING DATA ANALYSIS TECHNIQUES

With the advent of the personal computer and a wide range of software available for data analysis, all but the most simple surveys to a small group will benefit by converting the survey information into a machine readable form, or converting the words to numbers so that the computer can count data and generate calculations.

For simple analysis, you can use spreadsheets and relational databases. Social scientists commonly use software designed for advanced statistical analyses such as SAS, SPSS, SysStat, Minitab, and others. Many of these statistical analysis programs were originally mainframe-based, but most now have personal computer-based versions. The personal computer-based software will run on the 386, 486, and faster personal computers provided the computers have sufficient hard disk and temporary memory. While a math coprocessor is not absolutely necessary for such analyses, it will speed up data crunching.

Further advances in software development include modules for statistical programs that enable researchers to enter the questionnaire into the computer so that the data can be entered into what appears to be the questionnaire on screen. Such programs can also be programmed to warn data-entry clerks about potential errors in data entry. For a discussion of such techniques, see the manuals for SPSS and SAS.

Another advance is the use of computer coding sheets, or mark-sense sheets similar to those used for standardized testing in which the user marks the response in pencil on a machine-readable code sheet. The code sheets are then simply read into the computer. While such techniques work for simple questionnaires, developing automatic coding for more complex surveys is more challenging.

DEVELOPING THE CODEBOOK

Statistical analysis manuals offer two different techniques for assigning the data location in the data set: 1) fixed or 2) free format. For the fixed format, which is the method often preferred by many social scientists, a specific question is assigned to a specific column in the data set. For the free format, in contrast, you do not assign data to specific columns, but instead always place questions in the same order and use either blanks or commas to separate the data for your questions (SPSS 1988). For beginners, the fixed format minimizes the potential for errors.

To develop a codebook using the fixed format, you assign control numbers to each individual's questionnaire and specific columns to each question. Then you designate a numerical value to each possible response to each question.

The ethics of social science research dictate that you never enter the name of an individual in a database nor the social security number, student identification number, or other characteristic that would easily identify the respondent. In his classes on survey research, Professor Bud Sharp, who ran the University of Wisconsin's Survey Research Laboratory for more than 20 years, contended that he would go to jail to protect the confidentiality of an individual's responses. This confidentiality is as important as that of a doctor–patient or lawyer–client relationship. Respondents' unique responses are not revealed, and the aggregate data for the group provides confidentiality for the individual.

To protect each respondent, you should enter a control number on the questionnaire and keep these control numbers on a separate list that is not stored in the computer or with the questionnaires. Common safeguards include storing such lists and cover sheets (panel 12.2) in a separate room or facility under lock and key. Likewise, the names of individuals should never be reported in any document presenting the results of surveys.

The number of respondents sampled for your survey dictates the number of columns required for the control number. Thus a survey of less than 100 respondents requires a two-column control number—01 through 99; a survey of more than 100 respondents, but less than 1000, needs a three-column control number; and a survey of more than 1000 needs a four-column control number.

Next, you will assign specific columns to each question, and designate a numerical value to each possible response to each question. Consider the following questions from the survey used to survey research biologists of the Fish and Wildlife Service, U.S. Department of the Interior.

1. Are your publications primarily fisheries-related or primarily wildlife-related?

   ___ Primarily fisheries-related ___ Primarily wildlife-related

2. Does your office now submit manuscripts to the Office of Information Transfer?

   ___ Yes ___ No

2. a. If no, please explain why.

   ____

   ____

3. How many of your personal manuscripts has the Editorial Section edited during the last 24 months? _____

3. a. Of those manuscripts, how many were for Region 8 series publications?

   ____

3. b. Of those manuscripts, how many were nonseries publications—i.e., those manuscripts intended for journals and other scientific publications outside the Fish and Wildlife Service? ________

The following example represents the multiple deck strategy of coding for questions one through three.

Column		*Question #/Data Value*
1–3	ID number	Enter ID number, i.e., 001, 002, etc.
4	Deck	Enter 1
5–7	Date Recd	Enter day of month returned For column 5, let 1= January let 2= February For columns 6,7 enter day of month
8	Q1.What are your primary publications?	

1= fisheries

2= wildlife

9= not ascertained

9 Q2. Does your office now submit ms to OIT?

1= yes

2= no

10, 11 Q2a. No, Explain why?

01= Exempt

02= (NOTE: Reasons generated from responses, added during coding.)

03=

04=

05=

06=

07=

08=

09=

98= not applicable

99= not ascertained

12, 13 Q3. How many manuscripts did OIT edit last 24 months?

Code number

98= not applicable

99= not ascertained

14, 15 Q3a. How many Series 8 publications?

Code number

98 = not applicable

99 = not ascertained

16, 17 Q3b. How many were nonseries publications?

Code number

98 = not applicable

99 = not ascertained

Note that the codebook signifies the columns into which data are entered and also defines and identifies the data to be entered into the respective columns.

The researcher designated coded responses to question 1, which sought information about the subject of the respondent's publications, to be entered in column 8. For respondents who marked fisheries, the code "1" was designated and for respondents answering wildlife, the code "2" was designated. Likewise, for answers to question 2, a code of "1" was used for "yes" responses and a code of "2" for the "no" responses. If for some reason a respondent did not answer the question, a code of "9" was used. For 2a, an open-ended question asking why some offices might not be sending their manuscripts to the Fish and Wildlife's editorial office, two columns were allowed, creating the possibility of up to 99 different code categories for different responses.

The general approach to coding open-ended questions entails making out 3"x5" cards for each response. Each card contains the questionnaire control number, the column number, and the response. As you code each questionnaire, you make out the card. Once all questionnaires are coded, you sort the cards into major categories, assign codes to those categories, and then enter the appropriate code into the data set for each response in the correct column—in this case, columns 10 and 11. Note that "98" was designated *not applicable* and "99" *not ascertained.*

To code responses in question 3, the numbers were entered in columns 12 and 13. For entering from zero to nine publications, the responses were coded as 01 for one publication, 02 for two publications, through 09 for nine publications. For 10 through 97 publications, 10, 11, and so forth were entered for the respective number of publications.

Question 3a sought to ascertain the number of publications that were a specific kind of technical report known as a Series 8 publication with which the researchers were familiar. For this question, the researcher used the same coding strategy as for 3, and again used "98" for *not applicable* and "99" for *not ascertained* responses.

Researchers differ on how they code questions that are not applicable to an individual response or that were not ascertained. Some researchers leave the response category blank while others enter zeros, and others use 8s and 98s, or some other coding scheme. Likewise, researchers differ on how to handle data from these entries, and no single answer emerges. If you will need to break the responses into subcategories, you will need to use a coding scheme that will allow such manipulations of the data. You will also need computer programming to skip values such as 8s, 9s, 98s, and 99s. Some programs enable you to write a skip statement for several questions; others require individual skip statements for each question. Your consultation with a statistician, who is familiar with the software program that you plan to use, will be important. If you think that you might need advanced statistical analyses beyond frequencies and cross tabulations, consult with a statistician early in planning your codebook.

When creating a codebook, do not use every column. Instead, provide one, two, or three blank columns periodically at the ends of sections to the questionnaire or another logical breaking point. Having blank columns throughout the data set will avoid the problem of having to recode it entirely if you later find you made a mistake and need to expand a response from one to three columns. Furthermore, the blank columns provide a "visual" check of the data as you are entering it. Should you for some reason get off a column, the mis-entries would be easily spotted.

When developing codebooks for longer questions, you can save time and keyboarding efforts by using the block and copy commands of word processing systems and then editing the file as needed. For each subsequent codebook section detailing you could block and copy the section for columns 1–3 and 4, and then change the coding instructions to reflect a new file number in column 4. In fact, you can often speed building a codebook through the block and copy commands if the coding scheme is similar for different questions. For example, you can block and copy the designated codes for "yes" and "no" responses. When you do, keep your wits about you and pay attention to the details.

Once you have a working draft of a codebook developed, you need to carefully edit it for duplicate column numbers and to make sure that you have included coding schemes for all questions. Then pretest the codebook by entering data for half a dozen or more questionnaires to see if you encounter any problems. If all works well, you're ready to begin data entry.

ENTERING THE DATA

To enter data you can use a variety of software programs including statistical analysis programs; word processing, spreadsheet, and database programs; DOS editor programs; and data entry software.

Whenever you consider analyzing your own data, keep in mind that developing good data analysis skills takes time. While simple analyses may proceed with few problems, more complex problems can be time consuming and difficult. The following discussion provides a general overview that will be helpful whether you handle the process yourself or hire a statistician.

Universities often have a statistical consulting service, which may or may not be free. Often sociology, psychology, journalism, and communication departments have faculty members who are skilled in handling surveys and data analyses. Checking with either of these sources prior to data entry could preclude a variety of problems and pitfalls.

Statistical analysis programs allow data entry in two ways. Several programs allow you to enter the data directly into the program and create it as a file; other programs have subroutine programs for data entry. For a full discussion of data entry, see the respective software manuals for statistical packages.

The DOS editor, word processing, spreadsheet, and database programs provide an alternative way of entering data. With these programs, you can create the file as a regular word processing file simply by entering the data as you type. When you save the file, you save it as an ASCII or DOS text file that can then be imported into the respective software for the data analyses. Thus you need not have the statistical analysis software when entering data and conducting the analyses.

As you work, save the data frequently and make backup files. Whenever working on a substantial project, we suggest making three copies: one on the hard disk of the personal computer, and two on floppy disks. Store one set of backup disks at home or transfer the files to a computer in another building. In that way should some disaster strike—such as the failure of a hard disk or a floppy disk drive, fire, storm, earthquake, or flood—you can still recover the data.

As you enter the data, double-check your work to minimize errors in data entry. One way to check the data is to print the files and then examine each column for wild entries—entries that exceed the range of the codes for the respective column. You could also prepare the data analysis program and then run frequencies, or the percentage for the responses to the respective questions, and examine the data set for entries beyond the values. If these entries exist, you need to examine the coding for that question, identify the control number, retrieve the questionnaire, and correct the mistakes.

WRITING THE SOFTWARE PROGRAM

Before you write the software program, determine what kinds of statistics you need to run. The questions directing your project influence the statistical analyses. For most surveys, begin with frequency data and then progress to more sophisticated analyses using cross tabulations, chi-square, t tests, analyses of variance, regression, and other statistics depending upon the purpose of your survey, the nature of your research questions, and the kinds of data collected.

For many surveys, simple frequencies and cross tabulations will suffice. Frequencies are the percentage of responses to each question while cross tabulations entail comparing one question against several others. For example, in the survey of wildlife biologists, question 1 asked

Q1. What are your primary publications?

1= fisheries

2= wildlife

9= not ascertained

Since the survey explored whether fisheries or wildlife biologists saw the support of the Editorial Office differently, Q1 served as the basis for making the subsequent comparisons, or cross tabulations, by directing the computer to run a cross tabulation of Q1 with other selected questions.

Such cross tabulations allow you to develop a deeper understanding of the responses to questions. In contrast, straight frequencies may bury important information. It's much like looking at only the mean or average value and ignoring the distribution of responses as well as the shape of the distribution. Taking a closer look at such data can provide enhanced insights into the information being sought.

Once you have entered the data and selected the kinds of statistical analyses needed, you will need to instruct the software to run the desired analyses, or program the computer. A lengthy discussion of such programing is beyond the scope of this book, but a wide range of handbooks and manuals provide guidance for the respective statistical packages. The basic process involves writing a sequence of commands that tells the computer where to find the data, how the data are structured, the names of the variables, the variable location (columns that represent the different questions), the names of the variables, the names of the values, and the kind of analyses required.

Each software differs on the specifics, and some software has advantages over other systems depending upon the needed statistical analyses. Mostly likely, your initial needs for surveys will focus on generating frequencies, i.e., the percentages of different people responding to the different questions, and cross tabulations, i.e., comparing one group of respondents' answers to another group's on a specific question. All software programs will enable you to carry out these activities. In addition, the emerging Windows versions of software programs should enable users with few programming skills to run a wide variety of statistics. However, users who lack a clear understanding of statistics and their application to surveys and social science research should proceed with caution. Ignoring the assumptions underlying the statistics can lead to misuse of the statistics and erroneous conclusions.

RUNNING THE PROGRAM

While the details of running statistical analysis programs vary, the basic process entails logging onto the computer, loading the data files and your program, and then directing the computer to conduct the analyses. Specifics will depend upon how you set up the program, how you entered data, how the software was installed on the computer, and how you wrote the program to analyze the data. Again the manuals for the respective statistical packages provide these details. Likewise, study any specific instructions that may have been written for running the statistical packages on your particular computers.

For other computer facilities on university campuses, the computer support department often prepares a set of guidelines. For example, Colorado State University's Academic Computing and Networking Services (ACNS) produces a series of guides for using the software programs on the university's mainframe computers. The instructions cover such activities as how to log onto the computer, how to handle file transfers, and how to use the different statistical programs.

On many campuses, the computer support services and statistics departments may provide free technical support. At Colorado State University, for example, the ACNS provides walk-in consulting services. For assistance with statistical analyses, Colorado State's Department of Statistics offers walk-in support for troubleshooting problems with running specific statistical programs, and by-appointment consultation on planning the research design, sampling strategies, and selecting appropriate statistics. If you need additional help, try journalism, communication, sociology, and psychology departments; social research centers; and research laboratories at local universities and colleges. Faculty and staff of such facilities can provide a wide range of assistance depending upon their expertise and experience, which you should query.

If you are in a business, industry, or government agency, ask if any scientists and technicians have expertise in statistical analyses, using organization computers, and running respective programs. Such individuals also may be able to help you with your statistical analyses.

When you first attempt to conduct the statistical analyses, you will often have to debug the program. Errors creep in and you need to correct them before the program will run. Many software programs flag the errors, but it may take several runs to eliminate the problems. Occasionally you may need to turn to the computer manuals and consultants to help you solve the programming problem.

Once you have the bugs eliminated, and the frequencies run, examine the range of responses to each question. Look carefully to make sure that responses fall within the range expected. For example, you may have a question with only five response categories but the initial run produces responses in six.

Errors in typing and data coding often produce erroneous data points, so carefully examine the frequency responses for each category.

As you work, keep your wits about you, and carefully check the data files again. Run the frequencies again and check thoroughly to make sure that you have cleaned up any errors in the data and that you have not reintroduced new errors into the data set. In addition, examine the frequencies to make sure they are logically based on your coding of the original data. As you work through coding questionnaires, develop a sense of the frequency responses on key, or critical, questions. If the frequencies on the initial computer runs do not seem reasonable based on your review of the questionnaires, return to the questionnaires and count responses on key questions to make sure the coded frequencies are similar. If necessary, recheck the questionnaires, coding, and data sets.

Keep in mind that errors can easily creep in as you enter data and that having a "feel for the data" as you work through the survey, coding the questionnaires, and entering the data into the database, is critical in spotting possible errors. Simply, the data should make sense. Knowing the topic and having some sense or expectations of the results can help you avoid making serious errors.

Once you have the frequency data runs completed, you can move into making cross tabulations and running other statistics. The level of the statistical analyses required depends upon the purpose of the survey and the research questions guiding that survey. Chapter 16 provides additional guidance on selecting statistics and their applications, and a careful review of standard statistical textbooks will help you understand advanced applications. You might also review Kanji (1993), Kraemer and Thieman (1987), and Miller (1991).

Finally, a warning. Some statistical tests are quite easy to run, but most have several assumptions underlying them. To appropriately use them, you need to understand these underlying assumptions. If you do not, seek the advice of a researcher who does.

PRESESENTING YOUR FINDINGS

Once you have the statistical analyses completed, you are ready to begin interpreting the numbers. Steps include

- Reviewing research questions/objectives
- Selecting visuals
- Interpreting data and drawing conclusions
- Checking for common errors

Reviewing Research Questions

To begin, return to the original research questions that directed your project and consider each research question, problem statement, or objective. If you are writing a report or preparing a presentation, begin by listing each question separately. Consider which questions directly address your respective research questions/objectives and then consider what data you will need from the survey to answer the questions. In most cases, plan to summarize your information so that you can directly address your questions, present the necessary data, and draw conclusions. With that data in hand, turn to selecting the visual that will best present your findings to your audiences of readers, viewers, or listeners.

Selecting Visuals

When selecting a visual mode of data presentation—table, line graph, or bar graph—remember that readers must be skilled in reading and interpreting the particular data presentation to comprehend the data (Zimmerman and Clark 1987). Thus, you may find it necessary to use different data presentation techniques for different audiences.

Tables are most appropriate for sophisticated readers and listeners who have worked with data. If you are providing statistical analyses, consider whether or not your readers can understand statistical tests and the results. Remember, the general idea is to reduce the data set to a minimum number of tables or statistics from which you can draw your generalizations.

For less sophisticated audiences, and often for presentations, bar graphs, line graphs, and other visuals may be appropriate. For guidance on preparing visuals, see Anderson (1987), Enrick (1980), Horton (1991), or Zimmerman and Clark (1987). Keep in mind too that each scientific, technical, or specialized field has standard style guides that provide the specific formatting, organizing, and structuring details for preparing visuals.

Interpreting Data and Drawing Conclusions

Once you have your visuals prepared, try to summarize your findings in one or two key statements that clearly reflect the research question. To illustrate, consider the STC survey of technical communication writers, editors, managers, and freelancers to develop a clearer understanding of the technical communicator's roles and duties (Zimmerman, Muraski, and Peterson 1993). Table 13.1 summarizes the data designed to ascertain the range of tasks and their importance to technical communicators. Here's the researchers' interpretation of that data:

> The data clearly show that technical communicators must handle the production—planning, writing, and editing of a wide range of communication prod-

ucts—hardcopy (printed materials), online information, hypertext, magazine articles, market literature, videotapes, and slide sets (table 13.1).

Quite surprisingly, technical communicators are heavily involved in planning illustrations, preparing illustrations and working with illustrators, and they are likewise heavily involved in making presentations and training employees.

TABLE 13.1

THE FREQUENCY AND PERCENTAGE OF TECHNICAL WRITERS, EDITORS, AND MANAGERS ENGAGED IN PRODUCING SELECTED COMMUNICATION PRODUCTS (N=337) (ZIMMERMAN ET AL. 1993)*.

	Skill/ Knowledge	Percentage Answering
a. Planning projects	3.97	94 %
b. Coordinating projects	3.94	93
c. Researching projects	3.81	95
d. Interviewing	3.54	94
e. Organizing information	4.49	97
f. Writing hardcopy	4.04	91
g. Writing online copy or documentation	2.91	76
h. Writing for hypertext	1.60	42
i. Writing magazine articles	2.06	52
j. Writing market literature	2.40	59
k. Scripting videotapes	1.61	39
l. Scripting slide sets	1.69	43
m. Planning illustrations	3.36	80
n. Preparing illustrations	3.12	70
o. Working with illustrators	3.24	73
p. Conducting project reviews	3.16	73
q. Producing the document	4.05	90
r. Copy editing	4.06	93
s. Editing content—i.e., more than copy editing	4.22	96
t. Proofreading	3.93	95
u. Electronic publishing	3.99	81
v. Making presentations	2.55	82
w. Training employees	2.96	82
x. Testing what you produce	3.09	79

*Respondents rated the frequency of their use of the skill/activity on a 1 to 5 scale with 1 meaning "Hardly at all," and 5 being "Very often." The sampling error is ± 3.5%.

When interpreting data, strive to present the essential or key findings in a succinct statement that reflects the data and is supported by it. You do not need to discuss every point in the data, but you do need to interpret the findings for your readers. Remember to point out any unusual findings that need further exploration or investigation.

Checking for Common Errors

First, be aware that inexperienced researchers sometimes find themselves trapped by wide-ranging errors. As you work, avoid errors associated with technical content, hypothesis testing, level of your study, generalizations, and extrapolations. Check for technical errors on data handling, typing, keyboarding, and writing. Math and typographical errors are easy to make. Check data presentations against the original data sets, and carefully check the data you present and the narrative you develop. Have a colleague or editor review your copy, tables, visuals, and narrative for potential errors. Keep in mind that familiarity with the content often results in missing the obvious. So work carefully and seek help in checking your work.

Second, keep in mind that the scientific method never proves hypotheses, but can disprove the hypotheses (Kanji 1993). By disproving a hypothesis, you lend support to the alternative.

Third, consider your level of investigation. Surveys may provide descriptive data or correlation data about their subject. Therefore, unless you have used a survey to collect data for a controlled experiment or other sophisticated research design, you cannot make cause and effect statements based on survey data. For a discussion of research design, see chapter 16 as well as Campbell and Stanley (1963), Cook and Campbell (1979), Miller (1991), and Kerlinger (1986).

Fourth, avoid overgeneralizations. To begin, keep your sampling scheme in mind. If you used purposeful samples, talk only about your respondents rather than everyone in the potential population. If you sampled from a clean list of the population and you used a random sampling strategy, you can generalize back to the whole population and report the sampling error.

Fifth, avoid overextrapolation, or extending your interpretation beyond the data. Assume you were studying the level of recycling of office/waste paper on a university campus, and you found an annual increase of five tons for 1990, 1991, 1992, and 1993. You cannot project that the university will continue to increase its recycling by five tons annually for the next decade. By 1993 all offices already may have been recycling all of their waste paper. As you work, ask yourself, "Have I extended the data beyond their limits?" Know the limits of your data and the implications for limiting the inferences you draw.

Once again, remember to avoid common errors by working carefully, keeping your wits about you, doublechecking your work, and having others review your work. Such an approach helps minimize the chances that problems will emerge.

REFERENCES

Anderson, P. 1987. *Technical writing*. New York: Harcourt Brace Jovanovich.

Campbell, D. T., and J. C. Stanley. 1963. *Experimental and quasi-experimental designs for research*. Chicago: Rand McNally.

Cook, T. D., and D. T. Campell. 1979. *Quasi-experimentation*. Boston: Houghton Mifflin.

Enrick, N. L. 1980. *Handbook of effective graphic and tabular communication*. Huntington, NY: Robert E. Krieger.

Horton, W. 1991. *Illustrating computer documentation*. New York: Wiley.

Kanji, G. K. 1993. *100 statistical tests*. Newbury Park, CA: Sage.

Kerlinger, F. N. 1986. *Foundations of behavioral research*. New York: Holt, Rinehart & Winston.

Kraemer, H. C., and S. Thieman. 1987. *How many subjects?* Newbury Park, CA: Sage.

Miller, D. C. 1991. *Handbook of research design and social measurement*. Newbury Park, CA: Sage.

SPSS. 1988. *SPSS-X user's guide*. Chicago, IL: SPSS

Zimmerman, D. E. 1990. Survey results of the editorial services of the Office of Information Transfer, U.S. Fish and Wildlife Service, U.S. Department of the Interior. Technical report. Fort Collins, CO, Dept. of Technical Journalism, Colorado State Univ.

Zimmerman, D. E., and D. G. Clark. 1987. *The Random House guide to scientific and technical communication*. New York: Random House.

Zimmerman, D. E., M. Muraski, and J. Peterson. 1993. Who are we? A look at the technical communicator's role. Technical report. Fort Collins, CO, Center for Research on Writing and Communication Technologies, Colorado State Univ.

Exploring Advanced Research and Evaluation Methodologies

CHAPTER

Usability Testing—An Evaluation Technique

With the introduction of personal computers in the early 1980s, dozens of companies touted their computers, software, and instructions as "user friendly" without testing their claims. Gradually a few companies began evaluating how "friendly" users actually found their products. This process, evaluating how easy products are to use, is known as usability testing.

Over the last 15 years, technical communicators have used a wide range of usability tests, assessments, and evaluations. For example, Bethke (1983) outlined methods for surveying customers, Mills and Dye (1985) outlined a protocol analysis for videotaping subjects, and Warren (1988) focused on using readability formulas and case studies. Grice and Ridway (1989) provided an overview of usability evaluation from the technical communication perspective. They noted that modern documents must be easy to use by the intended audience as well as being complete and accurate.

While this chapter explores the usability of technical and scientific communications, evaluators and researchers over the last 50 years have also used usability testing in the areas of military hardware, telecommunications equipment, and information technology (Byerley, Barnard, and May 1993). A review of the wide-ranging usability literature shows that many terms are used to describe the usability testing: usability assessments, usability engineering, and usability evaluations.

In this chapter "usability testing" means the process of evaluating how well the intended users can interact with a technology to carry out assigned activities. Although that technology is often computers, computer software, and the accompanying instructions, usability testing can be done on any equipment, equipment instructions, manuals, publications, books, or instructions.

The following discussion introduces key concepts associated with usability testing, provides a process to follow, and then suggests additional literature to review.

KEY CONCEPTS

A good foundation for conducting usability testing begins by understanding the key concepts underlying the following questions:

- Why should you conduct usability testing?
- What are the pitfalls of usability testing?
- When do you conduct usability testing?
- Where do you conduct usability testing?
- What data collection methods do you use?
- What do you measure?

Why Should You Conduct Usability Testing?

The basic rationale underlying usability testing is making a better product or making instructions easier for people to use. Thus, usability testing can 1) make technical communication easier for people to use, 2) enhance the quality of productions, 3) promote greater user satisfaction, and 4) save both the producer and the users of the product money.

To illustrate, Moore, Eyre, and Rideout (1986) tested software installation procedures and cut the software installation time by more than 50 percent. Furthermore, all of their initial test subjects needed help but none of the final subjects did. Nielsen (1993) documented cost savings resulting from usability evaluations ranging from $40,000 to $500,000. Redish (1994) proposed a method of establishing the value of usability testing by 1) conducting usability testing with appropriate subjects, 2) recording the time required, 3) identifying any problems, 4) fixing the identified problems, and 5) retesting the subjects. Subtracting the time of the second test from the time of the first test determines the time savings. Multiplying the time savings by the number of potential users and their average hourly salary produces a potential cost savings generated from usability testing.

Major companies such as IBM, Hewlett Packard, Microsoft, and WordPerfect regularly conduct usability tests and evaluations of their equipment, software, manuals, instructions, and documentation, and more companies are turning to usability testing to improve their products and to help boost sales.

What Are the Pitfalls of Usability Testing?

Usability testing can produce potential savings and provide useful information; it can also lead to faulty conclusions. Such testing, like many other social science research methodologies, appears deceptively easy but is fraught with pitfalls. Holleran (1991) explores pitfalls associated with 1) sampling, 2) planning and conducting tests, 3) validity and reliability, and 4) misinterpretations of results.

The first pitfall, associated with sampling and subject selection, may preclude the generalizability of information collected. The problem centers around subject characteristics and sampling strategies. Clearly delineating the desired background of test subjects, however, is difficult because few product developers have a clear, detailed understanding of their intended users of the final product.

Assume you plan to conduct usability testing of a new word processing software for Windows. If you said test subjects should be familiar with Windows software, what exactly do you mean? Should they have experience only with the basic Windows interface or also with other Windows word processing software? How much experience should they have? One year, two years? Does one year mean working 40 hours a week using Windows word processing programs or working 20 hours a week? With what functions should they be familiar and how familiar would they be? The questions go on and on.

Subject selection also presents a problem from the statistical perspective. Since subject selection should not bias the potential responses, social science researchers routinely use random sampling of a known population. Random sampling ensures that each unit in the population has the same and equal probability of being selected. The characteristics of this limited number of individuals from the population can then be generalized back to the larger population. In contrast, most usability evaluations use purposively selected subjects, and therefore the results of the usability evaluation cannot be generalized. Evaluators conducting usability testing need to be especially cognizant of the limited generalizability of the findings and approach their findings with caution.

The second major pitfall of usability testing centers around methodological issues—i.e., the experimental design of the usability tests. Holleran documents potential problems of subject motivation and the demand characteristics of the usability tests. Holleran points out that the subjects may, in usability tests, perform tasks they normally would avoid, stay with tasks longer than they normally would, and attempt to solve problems in ways that they normally would not. Such factors threaten the validity of the usability measurements. Holleran also points out that the researchers/evaluators running the usability evaluation may inadvertently bias the results through information presented subjects, subject treatment, data handling, or data interpretation.

The third pitfall involves validity and reliability issues. Validity focuses on whether you are measuring what you think you are measuring and reliability focuses on the stability of your measurements over time (See chapters 1 and 16). Holleran identifies validity issues surrounding tasks presented to testers, self-report measures from questionnaires, protocols used, and subjects' abilities to articulate their thoughts, as well as the potential reactive effects of the usability test setting. Reliability issues surface when observers or coders must code or measure specific behaviors of users during the usability evaluations (Holleran). For example, what constitutes an "error" in using the software? Evaluators must write out a definition of an "error," have different coders observe the same usability sequence, and then compare the accuracy of their coding. In the same way, intercoder reliability means that a group of coders, examining the same videotape or making the same observation, should have a high level of agreement—i.e., their coding should have a high correlation (Wimmer and Dominick 1991).

The fourth pitfall involves interpretation of the findings of usability testing. Holleran cautions that individuals conducting usability testing may easily misinterpret, misunderstand, or misapply usability test results. According to Holleran, usability testing that generates only observation protocols, with no concrete information or no quantifiable data, may be faulted by subjective analysis and experimental bias. Quantitative data, or numbers, if properly collected, can be analyzed objectively and subjected to statistical analyses. To provide such statistical analyses, Holleran recommends content analysis (see Krippendorff 1980; Rosengren 1981; Weber 1990) and structural analysis (Bailey and Kay 1987).

While usability testing has potential pitfalls, carefully designed, executed, and interpreted usability testing can prove extremely beneficial to technical and professional communications. The stronger the evaluator's understanding of social science research methodologies and the more rigorous the design of the usability testing, the less likely the evaluator will be led astray by methodological issues and misinterpretation of data. Conducting effective usability evaluations requires careful attention to the details.

When Do You Conduct Usability Testing?

Over the last two decades, social scientists have developed evaluation research methodologies that have applicability to usability eval-uations. Generally, these evaluations are either formative evaluation or summative evaluation.

Formative evaluation, which occurs during the development of a product, service, or program, provides more insights and opportunities to correct problems. Conducting usability testing early in the development cycle and continuing testing and retesting throughout the project development often identifies problems or difficulties that can often be easily corrected. If left uncorrected, the problems may prove impossible to correct later. An iterative, or repetitive, approach to usability testing—design, test, correct, retest, redesign, and retest—provides continuous feedback to enhance a product before use.

Summative evaluations, which occur at the end of the product development, service, or program, can prove useful in identifying problems with existing products. The information gained from summative evaluations can then be used for revising an existing product for its next development release or for guiding the development of a new product.

Given the choice, a formative evaluation following the iterative approach provides the best approach. However, conducting usability testing anytime during development is better than not conducting any evaluation of the product.

Where Do You Conduct Usability Testing?

Usability testing can be conducted in either a laboratory or in the field.

Usability laboratories, usually modeled after testing facilities in psychology laboratories, often include at least two rooms—a test room and the observers' room—with a one-way mirror between the rooms. The test room includes a work area for the computer, hardware, or products to be tested. One or more video cameras and a microphone in the test room are connected to equipment in the observation room. The minimum equipment in the observation room usually includes a mixing panel that consists of a video monitor for each camera, an audio input unit, a date-time generator, and a mixer-dissolve unit.

The mixer-dissolve unit can mix the images from two cameras by superimposing a second image on the videotape. For example, you can capture subjects' facial reactions with one camera while watching the computer screen or the keystrokes or mouse movements. If the project involves computers, installing a video capture card enables evaluators to capture the visual image of the computer screen without using a camera. The date-time generator creates and sends a code that places the time and date of the videotaping to the VCR unit and records the information on the videotape. The microphone and receiver send a soundtrack of the test subject's voice to the VCR.

Depending on the available facilities, evaluators may use additional computers for logging information and more sophisticated cameras that are mounted on motorized tracks and have remote control lenses. The sophistication of a usability laboratory is limited only by the available funds for building and equipping it. In the early 1990s, a simple laboratory cost between $10,000-$15,000, while a fully equipped state-of-the-art usability laboratory could exceed $100,000.

Overall, a formal usability laboratory provides excellent control over the usability test arrangement, standardizes the test environment, minimizes threats to validity, and enhances the quality of data—if the usability testing is developed carefully and carried out with caution.

Usability testing can also be carried out in the field where the equipment and instructions are being used. For field testing, a single portable video camera with microphone and tripod can provide a basic record of a usability test. While such recordings can provide useful and beneficial information, they may not provide as much detail as usability tests run in a formal laboratory. However, conducting the usability tests in the field may prove necessary for some projects and can provide unexpected insights.

For example, Slater and colleagues (forthcoming) and Zimmerman, Slater, and Tipton (1992) conducted usability tests of a cancer-information, multimedia program in a center that provided training and education programs for low income women. Conducting the usability tests in the center had several advantages. The center had day care for the women's children and was an easily accessible, nonthreatening environment for the women. Further, the women trusted the center staff, who encouraged the women to volunteer as test subjects.

On-site usability tests of equipment, software, or instructions can give additional insights into their usability in real world settings. Evaluating existing software being used on the job will show how users may have adapted or changed the software to solve their specific or unique problems. The work environment may provide additional insights into the design of software or instructions, as when Zimmerman and Shimoda (1990) observed that white lettering on a blue background appeared as blue letters in a field usability test. General guidance in selecting colors for screen design suggests that designers avoid blue letters on a blue background.

A careful assessment of the problem being addressed should determine whether to conduct usability tests in the laboratory, in the field, or both. In any case, usability tests must be carefully conceptualized, planned, and executed to provide useful and beneficial information.

What Data Collection Methods Do You Use?

Evaluation research benefits through triangulation in data collection—i.e., using multiple data-gathering strategies. The basic tenet is that multiple methods of data collection will minimize potential errors and erroneous conclusions based on single methodologies.

Typically, usability testing has capitalized on protocol analysis techniques for data collection. The evaluator gives subjects a specific task to complete, asks subjects to talk aloud as they work through the task, videotapes the subjects, analyzes the data, and then interprets the results.

Other techniques can enhance usability evaluations. Direct observation during the protocol analysis helps identify problems and provides immediate feedback to the assigned tasks. Follow-up personal interviews and surveys, completed immediately after the usability testing, provide subjects the opportunity to summarize their reactions to the products. In some cases, follow-up interviews two weeks or so after the usability evaluations can provide a measure of their long-term reaction to the products.

Transcribing, or typing out, the narrative and information from the videotapes to conduct an analysis of the transcriptions follows the basic research strategies of Ericsson and Simon (1984). However, such transcribing may require more than four hours for every hour of video tape. Content analysis of the videotapes provides a viable alternative if the usability evaluation has clearly identified the variables or factors to be evaluated. By carefully defining the key variables to be observed and then developing a coding scheme, evaluators can code the videotapes directly and streamline the analysis process. Such data can provide insights into the more frequently recurring problems and identify the range of problems that potential users might encounter.

What Do You Measure?

In developing a conceptualization of the acceptability of computer systems, Nielsen (1993) identified five components of usability: 1) easy to learn, 2) efficient to use, 3) easy to remember, 4) few errors, and 5) subjectively pleasing. These five categories provide general guidance to evaluators in their development of a refined problem statement not only for computer usability, but also for a wide range of equipment, instructions, manuals, multimedia products, references, and information products. These concepts must be refined to look at specific tasks or activities. Evaluators must then carefully define specific ideas or concepts to be measured and develop appropriate ways of measuring the concept.

To illustrate, Zimmerman and Shimoda (1990) conducted a usability test of a computer system designed to collect data in scientific laboratories and to log the data directly into a database. The system consisted of a personal computer, a control unit of additional computer cards, printer, software, and a cabling unit, which enabled laboratory users to hook the system up to scientific test instrumentation. The primary purpose of the usability evaluation was to identify the problems that laboratory technicians might encounter when setting up the system and operating it. The ultimate objective was to provide a system that could be set up and running in less than two hours. The usability test consisted of identifying the problems that users encountered, correcting the problems, and then retesting the equipment. The researchers defined three levels of error severity. Level 1 errors were problems that test subjects confronted and solved in less than two minutes. Level 2 errors were problems that subjects confronted and solved within two to five minutes either by themselves or by calling a help line. Level 3 errors were problems that took subjects more than five minutes to solve by themselves or by calling a helpline (Zimmerman and Shimoda 1990).

Other evaluators define errors differently. For example, major companies now consider it a serious error in a system if the user must spend more than one minute solving the problem. Effective measurements requires carefully defining the concepts to be measured and then describing how those concepts will be measured during the usability evaluations.

A MODEL OF USABILITY TESTING

By approaching usability testing from a problem solving perspective, you break the process into a series of individual activities (fig. 14.1).

Develop the Research Question

Begin by developing a clearly defined research question that focuses on specific overall activities. First prepare the general research question and then conduct a task analysis that identifies a series of subquestions to explore.

Conduct a Task Analysis

With the research questions clearly articulated, analyze the tasks, or activities, that you would like the subjects to carry out or evaluate. Task analysis entails breaking a specific activity down into infinitely smaller tasks until you isolate a series of mental and physical steps required to carry out the designated task. For each activity evaluated, a task analysis will identify each step and the potential problems and errors that users might encounter. The level of detail required will depend upon the usability evaluation being conducted and the initial research question.

To illustrate, consider the different functions involved in word processing programs. A simple program would need to have functions for 1) creating a new file, 2) saving a new file, 3) retrieving old files, 4) editing files, 5) page formatting 6) printing, and so forth. Consider analyzing the editing function to include such tasks as 1) adding text, 2) deleting text, 3) moving text, 4) searching text, 5) sorting text, and so forth. Next consider the steps needed to move text from one location to another location in the file. The steps include both mental and physical activities.

1. Identify the text passage to be moved.
2. Highlight or block the text to be moved.
3. Direct the computer to cut (or move) the text.
4. Identify the new location for the text.
5. Indicate to the computer the location (usually by positioning the cursor).

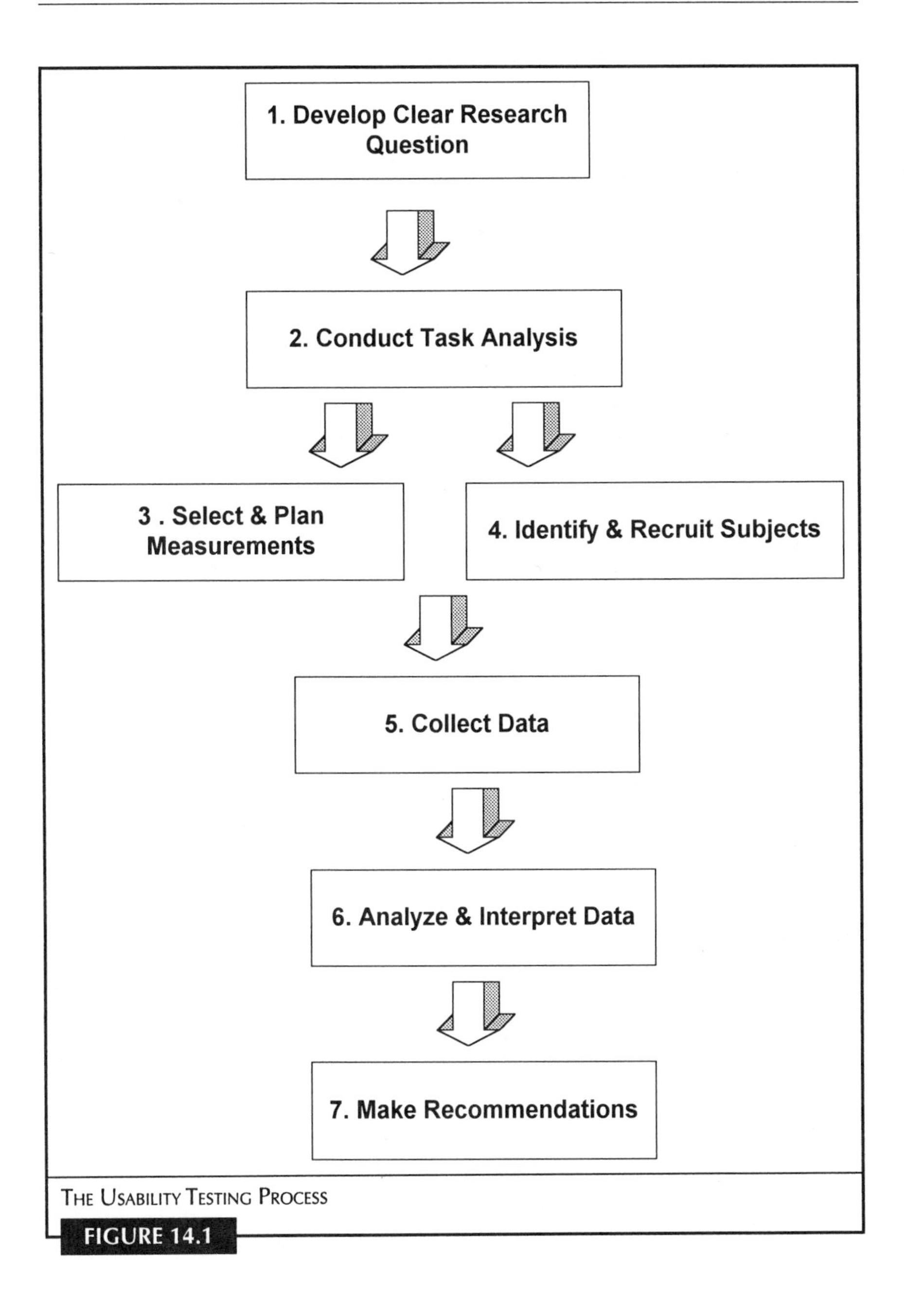

THE USABILITY TESTING PROCESS

FIGURE 14.1

6. Direct the computer to place the cut or moved text in the new location.

Task analysis emerged during World War II as a way of identifying information skills needed to operate specific military equipment and then developing a fast-track instruction on operating the equipment. To learn more about task analyses, see Zemke and Kramlinger (1987) and Jonassen, Hannum, and Tessmer (1989).

Select and Plan the Measurements

When planning a usability evaluation, consider using the following methodologies: 1) individual in-depth interviewing, 2) group interviewing, 3) surveys, and 4) protocol analysis. Earlier chapters discuss the first three techniques. Protocol analysis, a development of cognitive psychology, entails having subjects talk aloud as they think and work through an assigned task (Ericsson and Simon 1984). Always plan to use two or more measurement strategies to minimize the chances that a single method may lead you astray.

Preparing the measurements includes 1) writing a scenario, or set of instructions, for protocol analysis; 2) obtaining human subjects committee approval, as required; 3) arranging to use the usability laboratory or portable test equipment; 4) familiarizing yourself with the usability laboratory equipment and hardware as well as the software, hardware (equipment), and instructions or manuals to be evaluated; 5) preparing any interview schedules or questionnaires to be used after the protocol analyses; and 6) preparing check sheets to speed recording of the observations of test subjects.

Scenarios for protocol analyses give users a context for working and a clear set of directions or tasks to carry out. The scenarios thereby serve as context both for running the analyses and for evaluating the system being tested (Kyng 1992). Assume you were evaluating the ease of editing copy in a new word processing program. A possible scenario to evaluate the block, copy, and place functions of the software would be

> Assume you are learning XYZ word processing program and you need to move the last paragraph on page five to the top of page two. You are to use the mouse and the function keys to block copy the paragraph, scroll back three pages, and place the copied paragraph at the top of page two. Be sure to talk aloud as you work. If you encounter a problem, first try to solve it yourself using the "help" function and hardcopy manuals. If you cannot, then use the yellow telephone to call the helpline. When you have completed the task, please tell me.

When writing a scenario, include the following points: 1) set a stage or context for the protocol analysis, 2) explain the tasks to be carried out, and 3) give the subjects additional needed instructions. While developing the scenario carefully consider the characteristics of and the number of test subjects to be recruited, as will be discussed later in this chapter.

If you have access to a usability laboratory, make the needed arrangements to use the laboratory or video cameras for field tests. You can run a protocol analysis with one person, but having two or more people handle the details makes it easier. In industry, usability laboratories often use three or more staff members depending upon the evaluation being conducted and the facilities.

Familiarize yourself with the video equipment. You do not want problems with the video equipment to ruin the protocol analyses or to delay the testing of subjects. While many of today's cameras are easy to operate, some models have a dozen or more switches and buttons that can be adjusted. Read the instruction manuals and practice to make sure you can run the equipment easily.

Make sure that you have enough videotapes for the session. If possible, buy video tapes that are long enough to record all of each session on one tape. If the session will run longer than an hour, provide rest breaks for the subjects and time to change videotapes. If you will be running the video cameras on batteries, make sure the batteries are fully charged and you have backups. Keep in mind that equipment does fail. Be familiar enough with the equipment so that you can solve any problems quickly. Plan for at least half an hour to set up the equipment and half an hour to take down the equipment or to wrap up the usability tests after completing the protocol analyses.

If you plan to interview subjects either before or after the protocol analyses, prepare your questions or questionnaire to guide your interviewing. Keep in mind that you can ask subjects questions about their backgrounds, impression of the product, attitudes toward the problem, and specific reactions to the products.

To provide additional insights into the protocol analyses, plan to observe the test subjects as they carry out the tasks assigned in the scenario. Consider developing check sheets (panel 14.1) for recording your observations to provide both a standardized form and a way of quantifying the observations.

Finally, always plan to run a pretest of one or more subjects through the protocol analyses to identify any problems in the scenarios, equipment, or overall protocol procedures. Such pretests will also indicate the time needed to carry out specific tasks. In some cases, you may need to reduce the number of tasks to be evaluated.

Identify and Recruit Subjects

To generate useful information, you will need to determine the desired characteristics of test subjects, the number of subjects needed, how to identify the desired subjects, and how to recruit them.

To determine the desired characteristics of the test subjects, consider the ideal user—for whom is the product being developed? How will they be using the product being evaluated? What general background should the subjects have to easily use the product? What specific knowledge, skills, and expertise should the subjects have prior to using the product? Keep in mind that you can use the answers to such questions to help you recruit test subjects who fit the desired characteristics.

Consider again the proposed usability test for assessing a new word processing program. The user profile might specify that the intended users should be familiar with at least one of the leading word processing programs for Windows, such as Microsoft Word for Windows or WordPerfect for Windows. Their familiarity should include the ability to use all major editing functions including deletions, additions, copying, and moving. Prior to running the usability tests, the subjects should be able to demonstrate skills that include

PANEL 14.1

A CHECKLIST FOR OBSERVING TASKS IN PROTOCOL ANALYSES

Project Title: ______________________

Subject Number: ______________________

Date of Test: ______________________

Observer: ______________________

Directions: Write a succinct description of the task under "task" and enter the beginning and ending times. If a subject encounters a problem, note time when the problem was first encountered and when it was solved. Also note the strategies used to solve the problem.

Tasks	Time Begun	Time Ended	Problem Encountered	How Solved?
1.				
2.				
3.				
4.				
5.				
6.				
7.				
8.				

deleting individual words, sentences, paragraphs, and whole pages as well as blocking and copying one word, one sentence, and one paragraph within two minutes.

Once you have identified the desired characteristics of the subjects, consider how many subjects you will need for the usability tests. The number depends both on the methodology you will be using to collect data and the availability of subjects. See chapters 7, 8, 9, and 10 for guidance on the number of test subjects.

Nielsen (1993) provides a detailed statistical analysis of 36 published usability studies and reports that width of the confidence interval narrows moving from 3 to 25 subjects. As general guidance, Nielsen suggests adding subjects as long as the benefit from adding subjects exceeds the cost of running the usability test.

With a clear idea of the desired subjects, you can identify potential groups of individuals who might fit the desired characteristics. As with surveys, a list representative of the target population will provide the best method for selecting subjects. If such lists exist, you can randomly sample from them. If not, you may need to turn to purposeful sampling by seeking out individuals with the desired characteristics. For example, to run usability tests with experienced hypertext users, researchers recruited students from a technical writing course who had used hypertext applications (Bilsing 1994, Tipton 1994). While the results could not be generalized to all users, the researchers identified problems that experienced hypertext users had in navigating different hypertext structures.

Because usability testing requires one to two hours or more of a subject's time, most evaluators offer modest honorariums to the test subjects, as discussed earlier in regard to focus groups.

Do not forget that you may need to seek human subjects committee approval if you are conducting the usability tests on a college or university campus, as an employee of a government agency, or through government funded projects. In your application to the human subjects committee, you will need to explain how you will conduct the usability tests—e.g., protocol analyses, videotaping, interviewing, or other measurements. Keep in mind that such approval may take one to four weeks depending upon your organization and its policy of human subjects research and evaluation.

Collect Data

When subjects arrive to take part in the usability tests, give them a brief orientation to the project, but do not tell them so much that you sensitize them and bias the results. Strive to keep all orientation statements neutral and value free. If you must complete human subjects consent forms, first brief subjects on the specifics, then ask them to read, sign, and return the forms to you. Escort the test subjects to the laboratory, brief them on the equipment, read the scenario, and then give them a copy of the scenario.

During the usability tests, watch subjects carefully and collect direct observational data as they work. Be especially alert for subjects who may become frustrated if they cannot carry out the assigned tasks. Be prepared to promptly acknowledge that such subjects are doing a good job and that they have identified a major problem with the product or system being tested. Depending upon their overall reactions, consider stopping their testing. While you may encounter problems with an occasional subject, most subjects will not become frustrated in running the usability tests.

Immediately after the usability tests, conduct the interviews, if planned. Work quickly through the interview; since the test subjects have spent up to an hour or more on usability tests, they may be tired. If you noted any specific or unusual problems during usability tests, ask the test subjects about the problems they encountered and explore the potential reasons for the problems. Ascertain whether you expect the subjects to have such problems. If not, try to ascertain whether the problem is with the system or the test subject. As you close the interview, ask if you can check back in case you have additional questions as you analyze the data. Finally, acknowledge your sincere appreciation for the test subject's helping you evaluate the system.

After the subject leaves, review your notes, add additional comments, elaborate on specific problems encountered, and fill in any details needed to add insights and elaboration to the data collected. Review checklists, if used, and fill in the needed details. Keep in mind that you may forget details about the session, so document the details fully immediately after the session.

Analyze and Interpret Data

Immediately after a session write an overall assessment of the session—i.e., a subjective assessment of the system and each test subject's overall interaction with the system's strengths and weaknesses.

To begin the overall data analysis, review your subjective assessments, checklists, and interviews. Such a review can help you as you begin to quantify the data.

For the videotapes, consider using content analysis procedures to generate frequency data for the variables being investigated. Keep in mind that content analysis provides quantitative data, which gives you insights into the usability testing that subjective impression cannot. Further, quantitative data help evaluators avoid potential pitfalls of conducting usability testing. As a minimum, generate frequency data on the content analysis—i.e., the percentage of test subjects having success with the system and the percentage having problems with the system. Clearly identify the key problems with the system.

Analyze the data collected in in-depth interviews and surveys to give you even further insights into the usability tests. Follow the standard procedures explained in chapters 7, 8, and 13.

As you complete the data analysis, interpret the results. What do the findings mean to you? What are the primary problems that test subjects, as a group, had with the system being evaluated? Can you attribute the problems to the systems being evaluated or to your selection of test subjects? What other insights can you gain from the overall evaluation? What does it mean? Is the system appropriate for the intended users? Based on the findings, begin formulating recommendations for the system.

Make Recommendations

As you begin developing recommendations, consider whether you are conducting a formative or summative evaluation.

If a formative evaluation, what changes could be made in the current system to resolve the identified problems? Have you conducted the usability evaluation early enough in the system development cycle that changes can be made? If you cannot change a component of the system, can the technical communications—instructions, manuals, reference, tutorials, and online help—be revised to minimize the problems encountered? If changes can be made, will they minimize the problems encountered? Consider the need to use the iterative approach of redesign, retest, redesign until the problems are solved.

If the usability testing is a summative evaluation, the problems can be corrected for a subsequent version of the system. For example, software companies often release a new product and then within months announce an interim release. Such releases often correct problems, or bugs, in the original product.

When making recommendations, remember the limitations of the data collected. Avoid extending recommendations beyond the data presented unless you have designed the evaluation in such a way that you can generalize—i.e., used an experimental design with random sampling of test subjects as discussed in chapter 16.

ENHANCING USABILITY EVALUATIONS

The foregoing discussion provides a general orientation to usability testing. The prudent evaluator will strive to develop a deeper understanding of empirical social science research, communication research based on social sciences research methodologies, evaluation research, and human-computer interaction. The following discussion identifies selected helpful references.

A range of basic and advanced textbooks provides excellent discussions of social science research methodologies applicable to technical communication research. Babbie (1992); Williamson, Karp, Dalphin, and Gray (1982); and Nachmias and Nachmias (1987) provide a strong introduction to the basics of social science research. Kerlinger (1986) and Miller (1991) provide in-depth discussions of advanced social science research methodologies. Ericsson and Simon (1984) provide an in-depth discussion of protocol analysis as a research methodology for cognitive psychology research.

When researchers began exploring mass communication nearly three decades ago, they relied heavily on methodologies from sociology, psychology, and education. By the early 1990s, several authors had published textbooks on communication research based on a social science paradigm. They include Wimmer and Dominick (1991), Hsia (1988), Stempel and Westley (1988), and Singletary (1994). All provide broad-based overviews of key social science research concepts and methodologies with wide applicability to technical communication.

Evaluation research began emerging from the social sciences nearly two decades ago. Over the years, researchers have explored a wide range of social science research methodologies to solve day-to-day problems in organizations. A careful look at the work of Rossi and Freeman (1993); Scriven (1991); and Shadish, Cook, and Leviton (1991) can provide useful insights into evaluation strategies and techniques.

Most recently, researchers have written entire volumes on human computer interaction and usability. Booth (1989) provides a readable, detailed discussion of human-computer interaction and explores usability from that perspective. Preece (1993) provides an overview of usability as a tool for exploring human factors in computer use. Nielsen (1993) explores a range of applications for usability engineering in human-computer interaction. Byerley, Barnard, and May (1993) also provide an exploration of usability, communication, and computers, much of it from the European perspective. To keep abreast of new publications on usability testing, check *Books in Print*, research library catalogs, and book reviews in *Technical Communication*, *IEEE Transactions on Professional Communication*, and *Technical Communication Quarterly*.

Articles in journals and proceedings of national and international conferences are also excellent resources. Ramey (1989) edited a special edition of *IEEE Transactions on Professional Communication* that explored usability issues. Mirel (1991) provided a critical review of the usability research on hardcopy documentation; her work explored a series of methodological issues surrounding published usability studies. Duin (1993) provides overall guidance for hardcopy usability testing.

When considering usability testing, keep in mind that if handled improperly, it can lead to faulty conclusions. However, handled properly, usability testing can enhance the quality of many products by identifying problems during the production cycle.

REFERENCES

Babbie, E. 1992. *The practice of social research*. Belmont, CA: Wadsworth.

Bailey, W. A., and E. Kay. 1987. Structural analysis of verbal data. *Proceedings of Chi+GI*. Toronto, April 5–9.

Bethke, F. J. 1983. Measuring the usability of software manuals. *Technical Communication* 32(2): 13–16.

Bilsing, L. 1994. The role of external maps and spatial skills in reducing user disorientation in hypertext. Manuscript. Fort Collins, CO, Technical Journalism, Colorado State Univ.

Booth, P. 1989. *Introduction to human-computer interaction*. Hillsdale, NJ: Erlbaum.

Byerley, P. F., P. J. Barnard, and J. May, eds. 1993. *Computers, communication and usability*. New York: Elsevier.

Duin, A. H. 1993. Test drive: Evaluating the usability of documents. In *Techniques for technical communicators*, ed. C. M. Barnum and S. Carliner. New York: Macmillan.

Ericsson, K. A., and H. A. Simon. 1984. *Protocol analysis: Verbal reports as data*. Cambridge: MIT Press.

Grice, R. A., and L. S. Ridway. 1989. A discussion of modes and motives for usability evaluation. *IEEE Transactions on Professional Communication* 32(4): 230–37.

Holleran, P. A. 1991. A methodological note on pitfalls in usability testing. *Behaviour & Information Technology* 10(5): 345–57.

Hsia, H. J. 1988. *Mass communication research: A step-by-step approach*. Hillsdale, NJ: Erlbaum.

Jonassen, D. H., W. H. Hannum, and M. Tessmer, eds. 1989. *Handbook of task analysis procedures*. Englewood Cliffs, NJ: Educational Technology.

Kerlinger, F. N. 1986. *Foundations of behavioral research*. 3rd ed. New York: Holt, Rinehart & Winston.

Krippendorff, K. 1980. *Content analysis: An introduction to its methodology*. Newbury Park, CA: Sage.

Kyng, M. 1992. Scenario? Guilty. *SIGCHI Bulletin* 24(4): 8–9.

Miller, D. C. 1991. *Handbook of research design and social measurement*. Newbury Park, CA: Sage.

Mills, C. B., and K. L. Dye. 1985. Usability testing: User reviews. *Technical Communication* 32(4): 40–44.

Mirel, B. 1991. Critical review of experimental research on the usability of hard copy documentation. *IEEE Transactions on Professional Communication* 34(2): 109–22.

Moore, M. A. D., J. M. Eyre, and T. B. Rideout. 1986. Testing software installation procedures: An integrated approach. *Proceedings of the 33rd International Technical Communication Conference*. pp. 377–80.

Nachmias, D., and C. Nachmias. 1987. *Research methods in the social sciences*. New York: St. Martin's.

Nielsen, J. 1993. *Usability engineering*. Boston: AP Professional/Harcourt Brace Jovanovich.

Preece, J., ed. 1993. *A guide to usability*. Reading, MA: Addison-Wesley.

Ramey, J., ed. 1989. Usability testing. *IEEE Transactions on Professional Communication* 32(4): 205–316.

Redish, J. 1994. Usability testing: A critical aspect of T&D. *Total Quality Documentation* 2(1): 2–3, 4.

Rosengren, K. E., ed. 1981. *Advances in content analysis*. Newbury Park, CA: Sage.

Rossi, P. H., and H. E. Freeman. 1993. *Evaluation*. Newbury Park, CA: Sage.

Shadish, W. R., Jr., T. D. Cook, and L. C. Leviton. 1991. *Foundations of program evaluation*. Newbury Park, CA: Sage.

Scriven, M. 1991. *Evaluation thesaurus*. Newbury Park, CA: Sage.

Singletary, M. 1994. *Mass communication research*. New York: Longman.

Slater, M. D., D. E. Zimmerman, H. Halvorson, T. Kean, and J. D. Rost. (Forthcoming) Delivering health information to the disadvantaged: Assessing a hypertext approach. *Hypermedia*.

Stempel, G. H., and B. H. Westley. 1989. *Research methods in mass communication*. Englewood Cliffs, NJ: Prentice Hall.

Tipton, M. 1994. Navigating hierarchy and webs. Manuscript. Fort Collins, CO, Technical Journalism, Colorado State Univ.

Warren, T. 1988. Readers and microcomputers: Approaches to increase usability. *Technical Communication* 35(3): 188–92.

Weber, R. P. 1990. *Basic content analysis*. Newbury Park, CA: Sage.

Williamson, J. B., D. A. Karp, J. R. Dalphin, and P. S. Gray. 1982. *The research craft*. Boston: Little, Brown & Co.

Wimmer, R. D., and J. R. Dominick. 1991. *Mass media research*. Belmont, CA: Wadsworth.

Zemke, R., and T. Kramlinger. 1987. *Figuring things out: A trainer's guide to needs and task analysis*. Reading, MA: Addison-Wesley.

Zimmerman, D. E., and T. Shimoda. 1990. A usability assessment of the documentation and online instructions for the H-P System 10. Technical report for Hewlett Packard Instruments Division. Fort Collins, CO, Department of Technical Journalism, Colorado State Univ.

Zimmerman, D. E., M. Slater, and M. Tipton. 1992. Strategies for evaluating CD-ROM multimedia: Considerations for advancing communication effectiveness. *Proceedings of the STC Region 7* Conference. pp. 74–78.

CHAPTER

15

Ethnography and Case Studies

Underneath social science research lies human contact. A researcher collaborates with a subject to create a social relationship within which information is exchanged. Ethnography and case studies represent two approaches to these relationships.

Ethnography's distinguishing features include the elicitation of cultural knowledge (Spradley 1980), the detailed investigation of patterns of social interaction (Gumperz 1981), and a holistic analysis of societies (Lutz 1981). Ethnography is sometimes portrayed as descriptive research (Walker 1981) or, in contrast, as theory testing (Glaser and Strauss 1967; Denzin 1970).

Case studies, on the other hand, are

> 1) An analysis of a person or group, especially as a medical or social model; 2a) A study of a unit, such as a corporation, and the causes of its success or failure; 2b) An exemplary or cautionary model; an instructional example. (American Heritage College Dictionary 1993)

Wimmer and Dominick (1991) define a case study as one using as many data sources as possible to systematically investigate an individual, group, organization, or event. Yin (1989) distinguishes case studies as used for research from case studies used for teaching, from ethnographic and participant observation, and other qualitative methods. He defines a case study as "an empirical inquiry that investigates a contemporary phenomena within its real-life contents; when the boundaries between the phenomena and context are not clearly evident; and in which multiple sources of evidence are used."

This chapter will discuss both ethnography and case studies as approaches to data collection in technical communication research.

ETHNOGRAPHY DEFINED

Agar (1980) contends that the term "ethnography" represents both a process and a product. According to Agar, ethnography as a product is usually a book that presents historical, physical, biological, and social aspects of the group under examination. Fetterman (1989) describes the process of ethnography as the art and science of describing a group or culture. The act of "doing ethnography" is called fieldwork (Agar).

Fieldwork is the method of conducting long-term research to witness people and their behavior in real world settings. This naturalistic approach avoids the artificial response typical of controlled or laboratory conditions. Understanding the world—or some small fragment of it—requires that the ethnographer study it in all its complexity.

The field worker uses a variety of methods and techniques to ensure data integrity. These methods and techniques objectify and standardize the ethnographer's perceptions.

THE ETHNOGRAPHER'S METHODOLOGY

According to Fetterman (1989), before ethnographers begin asking the first question in the field, they must have the following in place.

- A problem statement
- A theory or model
- A research design
- Specific data collection techniques
- Tools for analysis

A Problem Statement

Fetterman (1989) offers that a researcher's orientation to a specific problem determines how a problem is defined. For example, psychologists, anthropologists, linguists, ethnographers, and technical communicators will define the concept of "reading comprehension" in radically different ways. Psychologists might define "reading comprehension" as an intrapersonal phenomenon, or "who" comprehends; anthropologists as a social phenomenon, or "when and where" comprehension occurs; linguists as a structural-linguistic phenomenon, or "what" is being comprehended; ethnographers as a socio-cultural phenomenon, or "how" varying degrees of comprehension in specific groups come about—including how cultural values are transmitted to create different values placed on reading comprehension; and technical communicators would most likely define "reading comprehension" as all of the above.

According to Fetterman (1989), the ethnographic research problem typically dictates the shape of the research design, including the budget, the tools to conduct the research, and even the presentation of research findings. The ethnographer's role further refines the definition of the problem, which usually reflects either a basic or an applied research orientation.

According to Hammersley and Atkinson (1983), the aim of the prefieldwork phase and the early stages of data collection is to turn the research problem into a set of questions to which a theoretical answer can be given, whether these questions are in the form of a narrative description of a sequence of events, a generalized account of the perspectives and practices of a particular group, or a more abstract theoretical formulation. Often, as a result of this process, original problems are either transformed or completely abandoned in favor of others.

The research problem also shapes the selection of a place and a people or program to study. According to Fetterman (1989), most ethnographers initially use the big net approach for participant observation; they mix and mingle with everyone they can. As the study progresses, the focus narrows to specific portions of the population under study. This approach ensures a wide-angle view before the microscopic study of specific interactions begins. In ethnography, sampling from the data available is important in making decisions about where to observe and when, whom to talk to and what to ask, as well as what to record and how. According to Hammersley and Atkinson, the criteria employed in these choices should be as explicit and as systematic as possible, so as to ensure that the case has been adequately sampled.

A Theory or Model

The ethnographer's theoretical approach—whether it is an explicit theory or an implicit personal model about how things work—helps define the problem and how to tackle it. According to Fetterman (1989), most ethnographers use either an ideational theory or a materialistic theory. Ideational theories suggest that fundamental change is the result of mental activity—thoughts and ideas (Fetterman 1989). Classic ideational theories include sociolinguistics and symbolic interactionism. In contrast, materialistic theories suggest that fundamental change is the result of material conditions—ecological resources, money, or modes of production (Fetterman 1989). Classic materialistic theories include historical materialism, or neo-Marxism, and cultural ecology.

Fetterman (1989) recommends that the selection of a theory should depend on its appropriateness, ease of use, and explanatory power. (For a more detailed discussion of theory in ethnographic research, see Dorr-Bremme 1985; Fetterman 1986; Simon 1986; and Studstill 1986).

A Research Design

Fieldwork, the most characteristic element of any ethnographic research design, shapes the research design of all ethnographic work. The following important concepts guide an ethnographer in fieldwork.

- Culture
- Holistic perspective
- Nonjudgmental orientation
- Micro- or macro-study

However defined, the concept of culture helps the ethnographer search for a logical, cohesive pattern—often ritualistic behaviors and ideas that characterize a group. After finding this pattern, the ethnographer may offer a cultural interpretation. Cultural interpretation describes what the researcher has heard and seen within the framework of the group's view of reality and rests on a foundation of carefully collected ethnographic data.

This cultural description may include the group's history, religion, politics, economy, and environment. Ethnographers must assume a holistic outlook in research to gain a comprehensive and complete picture of a social group. The holistic approach forces the field worker to see the cultural implications beyond an immediate individual scene or event.

While some ethnographic concepts push the researcher to explore in new directions, others ensure that the data are valid, and others simply prevent contamination of the data. A nonjudgmental orientation helps the ethnographer in all three ways. A nonjudgmental orientation requires the ethnographer to view another culture without making value judgments about unfamiliar practices.

An ethnographer's theoretical disposition and research problem determines whether the ethnographer conducts a micro- or macro-study. A micro-study is a close-up view of a small social unit or an identifiable activity within a social unit. A macro-study focuses on the large picture. However, both require the same amount of time to conduct.

Specific Data Collection Techniques

Ethnographers use the following data collection techniques.

- Participant observation
- Interviews
- Projective techniques

Participant observation characterizes most ethnographic research and, according to Fetterman (1989), is crucial to effective fieldwork. Participant observation combines participation in the lives of the people under study with a professional distance that allows observation and recording of data. Powdermaker's *Stranger and Friend* (1966) depicts this role. Participant observation sets the stage for more refined data collection—including interviews and projective techniques—and becomes more refined itself as the field worker understands more and more about the group under study. Participant observation can also help clarify the results of more refined instruments by providing a baseline of meaning and a way to explore the context of those results.

Interviews, the ethnographer's most important data gathering technique, explain and contextualize the fieldwork. According to Fetterman (1989), ethnographers extract every word's cultural or subcultural connotations as well as its denotative meaning. Interviews can be structured, semistructured, informal, and retrospective interviews. Each kind of interview seeks to solicit different information. For additional discussion on interviewing, see chapters 7 and 8, Taylor and Bogdan (1984), and Werner and Schoepfle (1987).

Projective techniques elicit cultural and psychological information from group members through their interpretation of images, dreams, pictures, and objects. For example, an ethnographer holds up an item and asks the group what it is. What the item actually is is less important than the participants' perceptions. The participants' responses usually reveal individual needs, fears, and inclinations.

Tools for Analysis

An ethnographer's analysis tools may include

- Triangulation
- Operationalism
- Content analysis
- Other methods

Triangulation is a method by which the ethnographer tests one source of information against another to narrow alternative explanations and prove a hypothesis. In data source comparisons, for example, the ethnographer compares information and the sources of information to understand more completely the whole situation. Hammersley and Atkinson (1983) also identify the triangulation that exists between different researchers when the data generated by different ethnographers are compared and contrasted, and the triangulation that occurs between techniques when ethnographers compare data produced by different techniques.

The concept of "operationalism" is important in the ethnographer's post-field phase. Operationalism means defining one's terms and method of measurement. In simple descriptive accounts, saying that "a few people said this and a few others said that" may not be problematic. However, establishing a significant relationship between facts and theory requires greater specificity. For example, the statement "Learning decreases when too many students are in class" may be an accurate observation. However, what constitutes learning? How is an increase of learning measured? How many are too many students in class? Instead of leaving conclusions to strong impressions, the field worker quantifies or identifies the source of ethnographic insights.

Ethnographers analyze written data in the same way they analyze observed behavior. They attempt to discover patterns within text and seek key events recorded in print or electronic media. For guidance on the qualitative approaches to content analysis, see Krippendorff (1980), Rosengren (1981), and Weber (1990). Computerized approaches and creating databases of information allow the researchers to sort, compare, contrast, aggregate, and synthesize data to explore such patterns.

In documenting approaches to data analysis, Fetterman (1989) suggests that researchers consider using maps, flow charts, organizational diagrams, and matrices to organize and explore their data. Generally, he recommends nonparametric statistics but cautions that sophisticated statistical analyses may be inappropriate for ethnographic research because of nonrandom selection of subjects, small sample sizes, and the possible violation of other assumptions underlying advanced statistical methods.

A TECHNICAL COMMUNICATION EXAMPLE

To develop a deeper understanding of technical and professional communication, Paradis, Dobrin, and Miller (1985) explored the writing practices in a research and development unit at Exxon Chemicals. The researchers set out to determine the roles of writing and its associated activities in a research and development organization and how engineers and scientists produce internal documents. Prior to the on-site data-gathering, Paradis, Dobrin, and Miller asked the 33 engineers and scientists to complete a questionnaire and writing test. During a week-long visit the researchers interviewed 26 professionals, interviewed small working groups, met with supervisors and managers, observed patterns, and inspected written documents. Such a project can be considered both an ethnographic study and a case study. For a detailed review of the findings, see Paradis, Dobrin, and Miller (1985).

As with all advanced research methodologies, ethnography can provide added insights into a variety of technical communication, as well as scientific and technical, topics that involve people. The cautious researchers considering ethnography should develop a deeper understanding of it through a careful study.

A CLOSER LOOK AT CASE STUDY RESEARCH

Case studies represent an almost universal approach to investigating aspects of a discipline, such as anthropology, business, education, engineering, communication, law, library science, medicine, or zoology. To illustrate, Wickman (1978) studied the impact of Douglas-fir tussock moth outbreak and subsequent timber conditions. Bissex and Bullock (1987) explored case-study research by writing instructors. Johnson (1984) investigated cooperative science between a national university and industry researchers.

The Harvard Business School and School of Education are well-known for their case studies. Journalists regularly use case studies (Yin 1989, Wimmer and Dominick 1991, and Hsia 1988), as do legal scholars (Hsia).

The following discussion reviews selected characteristics of case studies, proposes a generalized approach to conducting case studies, and provides suggestions for further research on case studies.

Case study methodologies are primarily developed from social science research methods, as detailed in textbooks and monographs devoted to specific topics. Researchers in other areas may also rely heavily on social science methodologies.

Historically, social science researchers who used quantitative approaches—i.e., number crunching—and researchers following qualitative approaches—i.e., a more verbal, descriptive approach—have argued about the value of their different methodologies. Quantitative researchers often criticize case study research because of its lack of rigor, lack of generalizability, possible subjective influence of the researcher, and lack of methods to reconcile contradictory findings across case studies (Hsia 1988; Wimmer and Dominick 1991).

However, in writing the foreword to Yin's *Case Study Research* (1984), Donald Campbell, a leading quantitative researcher, praised the case study methodology that Yin proposed and endorsed the need for case study research in social science research methods courses. Yin, a leading researcher in case study methodologies, explores the differences of using experiments, surveys, archival analyses, history, and case studies. He argues that case studies provide the opportunity to answer the "why" and "how" that some other social science methodologies may not be able to answer (1989). As discussed earlier, triangulation—using multiple data gathering strategies—minimizes the possibilities of erroneous data collection.

Most researchers divide case studies into the individual case study or multiple case study. An individual case study concentrates on one individual, unit, organization, or group. A multiple case study explores two or more similar individuals, units, organizations, or groups.

Works that explore various kinds of case studies include Hsia (1988), Singletary (1993), Towl (1969), Williamson and colleagues (1982), and Wimmer and Dominick (1991). Beyond studying specific studies, Wimmer and Dominick stress that case studies help people to understand the topic being studied and that case studies depend heavily on inductive reasoning based on the generalizations and principles that emerge from the data.

A Case Study Process

Wimmer and Dominick propose a case study methodology that has broad applicability to many disciplines and can be a foundation for building case study research methodology skills. They suggest five steps for a case study (fig. 15.1).

Designing the Case Study. Like all research, case studies begin with a clearly defined question. Research questions guiding case studies generally focus on exploring the "why" and "how" dimensions of the problem being addressed. A careful literature review to learn how others have explored the topic will provide added insights and help develop more effective and targeted research questions. As with all problem solving, developing and articulating a research question helps shape the subsequent investigation and clearly identifies the data needed to address that question.

With the research question developed, determine whether you will study one or multiple units. When first using a case study methodology, an individual case study will help you develop your research skills. Be careful not to undertake a project that cannot be completed within the available time. Decide the limits of your case study. Wimmer and Dominick (1991) suggest that your literature review can help define those limits.

Developing the Pilot Study. Before conducting the pilot study, Wimmer and Dominick (1991) suggest that you first develop a protocol—a detailed plan that outlines the data-collection methodologies, a timetable, and the resources needed to collect the data. Traditionally, data collection methods for case studies have concentrated on 1) existing documents to be reviewed, studied, and examined, 2) interviews, 3) observation/participation, and 4) physical artifacts.

For some case studies, the existing documents may, in fact, be physical artifacts; in other cases artifacts may have unique physical properties. For example, artifacts in a case study of a particular forest insect may be a count of the percentage of dead trees in a given area. In a case study on the writing skills of engineers, the artifacts might be the various files (outlines and drafts) developed in preparing proposals or writing reports.

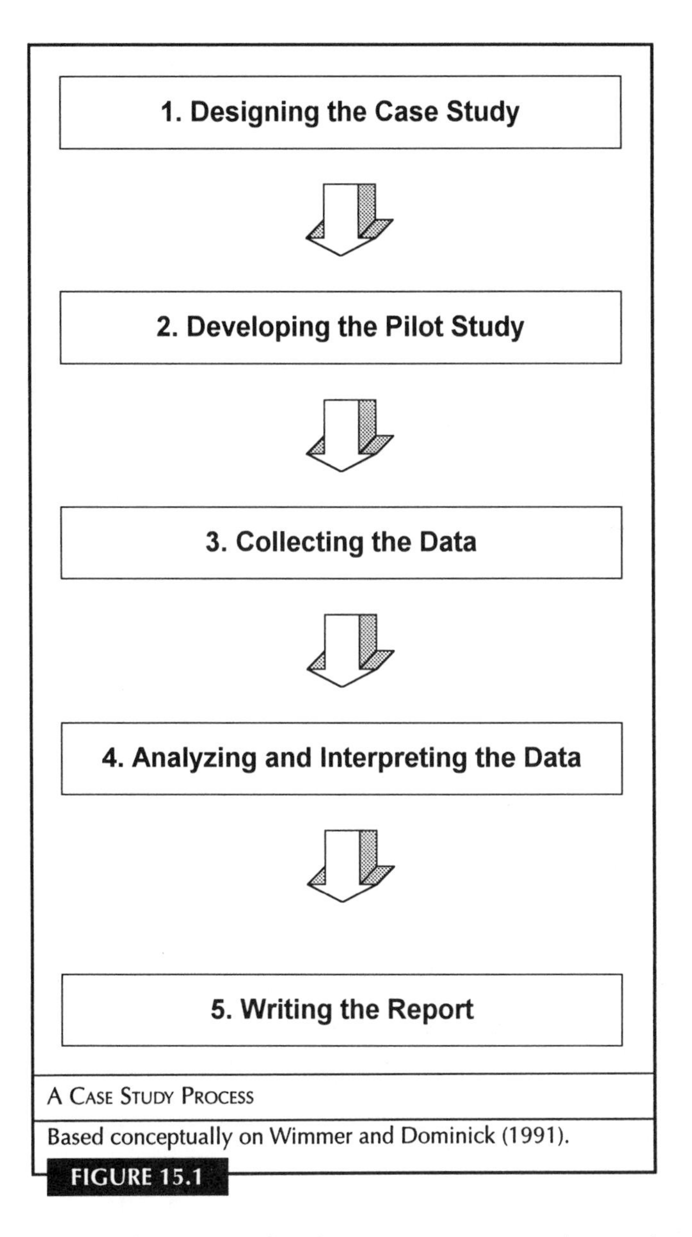

A CASE STUDY PROCESS

Based conceptually on Wimmer and Dominick (1991).

FIGURE 15.1

Nothing precludes a researcher from using a variety of research methodologies to collect relevant data. In fact, triangulation of data minimizes the chances of drawing erroneous conclusions due to errors in a particular data collecting methodology. Multiple methods thus help ensure a valid and reliable study.

In developing the timetable, leave adequate time to collect and analyze the data before writing the report. This may be difficult for your first case study. Break all the tasks down into subactivities and base your time estimates on 1) related experience you have had, 2) guidance required for different methodologies discussed in earlier chapters and the research cited, and 3) advice from researchers experienced in the various methodologies that you will use. Keep in mind too that most projects take more time than most researchers initially estimate. Some researchers add a cushion to their best estimates, sometimes even doubling those estimates. As you work through the case study, keep records of the time required for different activities. This will help you generate more realistic time estimates for your next case study.

As you select the methodologies that you plan to use and project the time, begin identifying required resources to carry out the respective project. For example, will a copy machine be available for photocopying records and documents? Would laptop computers help record data? What kind of budget will be needed to cover the overall expenses? Out-of-pocket expenses? Will you need to transcribe the complete narrative of the interviews?

Collecting the Data. As you collect the data, follow the suggested procedures discussed in the chapters on literature reviewing, interviewing, surveying, ethnography/participant observation, and usability. Keep in mind that each methodology requires a clear research question and careful attention to details in the respective methodology. As you work, keep careful records to preclude problems as you analyze and interpret the data.

Analyzing and Interpreting the Data. Two general analytic strategies for data analysis and interpretation are the theoretical approach and the descriptive approach (Yin 1989).

The theoretical approach allows the theory directing the case study to direct your analysis. The advantage of using a theoretical approach is that the theory helps you develop the research question, and the research question clearly directs the data analysis and interpretation—i.e., it puts limits on the data and subsequent interpretation.

The descriptive approach needs a framework for analyzing the data. Such an approach is more difficult because you need to develop a way of organizing the information so that you can then organize the analysis. It's much like organizing a general literature review report. While several organizations are plausible, you need to pick the best one to fit your research question.

To illustrate, assume you were to conduct a case study of the writing processes of university professors. Your research question focused on how the writing processes vary among professors. Assume that you interviewed professors in a variety of fields and that you collected samples of the materals—notes, outlines, and drafts—both in hardcopy and computer word processing files. At this point, you could choose descriptive approaches focusing on 1) the major written products that the respective professors produce, 2) the processes that professors use when writing, or 3) the writing tools—paper and pen, typewriters, and word processors—that professors use.

Further, your data analysis and interpretation will depend upon the data collection techniques used. For quantitative research methods, follow the guide in the respective chapters covering the methodology. Depending upon the research questions posed, percentages and statistical tests may be used to describe the data.

For qualitative approaches—interviews, ethnographic participant observation, and publication analysis—Yin (1989) recommends pattern matching, explanation building, and time-series analysis. Briefly, a simple pattern matching analysis compares the theoretical prediction with the data collected for the case study. Other pattern-matching strategies include nonequivalent dependent variables and rival explanations. For a detailed discussion, see Yin.

Explanation building entails making causal statements, or links (Yin). Researchers using explanation building write a narrative describing the case and linking the main concepts or ideas. General explanations then emerge from the narrative.

Time series analysis, a complex technique, follows approaches used in experiments and quasi-experiments that track subjects or units over time (Yin). Such strategies are best left to advanced research methodologies. For a detailed discussion, see Yin.

Writing the Report. Writing reports on case studies centers around organizational strategies and the intended audience.

Some case studies may lend themselves well to the general introduction-methods-findings-discussion organization, other studies to common organizational patterns of literature reviews, such as 1) general to particular, 2) particular to general, 3) pattern arrangement, 4) functional or process oriented, 5) order of importance, 6) chronological or historical, 7) cause and effect, or 8) combinations thereof (Zimmerman and Rodrigues 1992). Still other studies may lend themselves especially well to the problem-solving and design report organization pattern that 1) identifies the problem, 2) explains the potential magnitude of the problem, 3) narrows the solution to the one(s) proposed, 4) explains the problem-solving processes used, and 5) documents the final outcome (Zimmerman and Rodrigues).

While the organization is critical to writing the case study report, the intended audience is also influential—not only on the organization, but also on the language. When writing for experts in the respective fields, you may not need to explain terms and terminology as you would for a more general audience.

For specifics on writing research reports, Zimmerman and Rodrigues provide overall guidance on organizing, drafting, revising, and finalizing reports.

ADVANCED CASE STUDY RESEARCH

For more information on case studies, we recommend Hamel (1993), Hammond (1964), and Yin (1989, 1993) for general guidance on conducting case studies. For information relevant to specific disciplines, a careful literature review should provide information on case study methodologies in the respective disciplines.

REFERENCES

Agar, M. H. 1980. *The professional stranger: An informal introduction to ethnography.* Orlando, FL: Academic Press.

Bissex, G. L., and R. H. Bullock. 1987. *Seeing for ourselves: Case study research by teachers of writing.* Portsmouth, NH: Heinemann.

Denzin. 1970. *The research act: A theoretical introduction to sociological methods.* Chicago: Aleine.

Dorr-Bremme, D. W. 1985. Ethnographic evaluation: A theory and method. *Educational Evaluation and Policy Analysis* 7(1): 65–83.

Fetterman, D. M. 1989. *Ethnography: Step by step.* Newbury Park, CA: Sage.

Fetterman, D. M. 1986. Beyond the status quo in ethnographic educational evaluation. In *Educational evaluation: Ethnography in theory, practice, and politics.* eds. D.M. Fetterman, and M.A. Pitman. Newbury Park, CA: Sage.

Glaser, B., and A. Strauss. 1967. *The discovery of grounded theory.* Chicago: Aldine.

Gumperz, J. 1981. Conversational inference and classroom learning. In *Ethnography and language in educational settings.* eds. J.L. Green, and C. Wallat. Norwood, NJ:Ablex.

Hamel, J. 1993. *Case study methods.* Newbury Park, CA: Sage.

Hammond, P. E. 1968. *Sociologists at work: Essays on the craft of social research.* Garden City, NY: Doubleday.

Hammersley, M., and P. Atkinson. 1983. *Ethnography: Principles in practice.* New York: Tavistock.

Hominid, P. E., ed. 1964. *Sociologists at work.* New York: Basic Books.

Hsia, H.J. 1988. *Mass communications research: A step by step approach.* Hillsdale, NJ: Erlbaum.

Johnson, E. C. 1984. *Cooperative science: A national study of university and industry researchers.* Washington, DC: Productivity Improvement Research Section, Division of Industrial Science and Technological Innovation, National Science Foundation.

Krippendorff, K. 1980. *Content analysis: An introduction to its methodology.* Newbury Park, CA: Sage.

Lutz, F.W. 1981. Ethnography—The holistic approach to understanding schooling. In *Ethnography and language in educational settings,* eds. J.L. Green and C. Wallat. Norwood, NJ:Ablex.

Paradis, J., D. Dobrin, and R. Miller. 1985. Writing at EXXON ITD: Notes on the writing environment of an R&D organization. In *Writing in nonacademic setting*, eds. L. Odell and D. Goswami. New York: Guilford.

Powdermaker, H. 1966. *Stranger and friend: The way of an anthropologist*. New York: Norton.

Rosengren, K. E. ed. 1981. *Advances in content analysis*. Newbury Park, CA: Sage.

Simon, E.L. 1986. Theory in educational evaluation: Or, what's wrong with generic-brand anthropology. In *Educational evaluation: Ethnography in theory, practice, and politics*, eds. D.M. Fetterman, and M.A. Pitman. Newbury Park, CA: Sage.

Singletary, M. 1993. *Mass communication research*. New York: Longman.

Spradley, J. P. 1980. *Participant observation*. New York: Holt, Rinehart & Winston.

Studstill, J.D. 1986. Attribution in Zairian secondary schools: Ethnographic evaluation and sociocultural systems. In *Educational evaluation: Ethnography in theory, practice, and politics*, eds. D.M. Fetterman, and M.A. Pitman. Newbury Park, CA: Sage.

Taylor, S. J., and R. Bogdan. 1984. *Introduction to qualitative research methods: The search for meanings*. New York: Wiley.

Towl, A. R. 1969. *To study administrations by cases*. Boston: Harvard University Business School.

Walker, R. 1981. On the uses of fiction in educational research. In *Practising evaluation*, ed. D. Smetherham. Driffield, Eng.: Nafferton.

Weber, R. P. 1990. *Basic content analysis*. Newbury Park, CA: Sage.

Werner, O., and G. M. Schoepfle. 1987. *Systematic fieldwork*. Newbury Park, CA: Sage.

Wickman, B. E. 1978. *A case study of Douglas-fir tussock moth outbreak and stand conditions 10 years later*. Portland, OR: Dept. of Agriculture, Forest Service, Pacific Northwest Experiment Station.

Williamson, J. B., D. A. Karp, J. R. Dalphin, P. S. Gray. 1982. *The research craft*. Boston, MA: Little, Brown & Co.

Wimmer, R. D., and J. R. Dominick. 1991. *Mass media research*. Belmont, CA: Wadsworth.

Yin, R.K. 1993. *Applications of case study research*. Newbury Park, CA: Sage.

Yin, R. K. 1989. *Case study research: Design and methods*. 2d ed. Newbury Park, CA: Sage.

Yin, R. K. 1984. *Case study research: Design and methods*. Newbury Park, CA: Sage.

Zimmerman, D. E., and D. Rodrigues. 1992. *Research and writing in the disciplines*. Fort Worth, TX: Harcourt, Brace, Jovanovich.

CHAPTER

16

Exploring Advanced Research Methodologies

This chapter identifies advanced research techniques, explores key concepts underlying advanced research methodologies, and identifies relevant references for further study.

SCIENTIFIC STUDIES

Scientific studies can be classified as descriptive, associative, or cause and effect studies. In descriptive studies researchers collect data and describe what they found. In associative studies researchers collect data and report relationships, often correlations, among the variables or concepts being studied. In cause and effect studies researchers use experimental and quasi-experimental design—i.e., treatment and control groups, whereby one group receives the treatment and the other does not—to collect data. The researchers then compare the two groups on selected characteristics and make cause and effect inferences.

One research project may include descriptive, associative, and cause and effect studies. For example, in fall 1993 communication researchers at Colorado State University began a multiphase research project designed to enhance the communication skills of electrical engineering graduates. The researchers (Zimmerman et al. 1994) began first with a series of descriptive and associative studies.

The researchers collected data for their descriptive studies using surveys and in-depth interviews of electrical engineering majors, professors, recent graduates, and supervisors of recent graduates.

These same methods also can provide associative data. For example, the survey of electrical engineering majors not only explored their communication skills, but also asked questions about their attitudes toward computers.

Through statistical analysis, the researchers then ran correlations between the questions assessing students' attitudes toward computers, their use of computers in general, and their use of word processing programs.

As part of the multiphase project, the researchers are developing multimedia modules (hypertext programs) to help electrical engineering students write a variety of assignments. For example, these modules would provide the overall structure of a complete laboratory report, give guidance on how to write different parts of the report, tell how other students prepared the assignment, detail selected problems that other students encountered, and provide writing tips.

Thus, researchers could conduct an experiment to answer the question, "Does providing access to multimedia modules help students improve the quality of their laboratory reports?" They would assign students to two groups. One group would have access to the multimedia modules while the other group would not. The researchers would then measure, or evaluate, the laboratory reports of students from both groups. If the students using the multimedia modules prepared better laboratory reports, the researchers could then conclude that the modules appear to enhance the quality of electrical engineering majors' laboratory reports.

The key to the researchers' being able to infer that multimedia modules helped students prepare laboratory reports lies in using an experimental design. Researchers planning cause and effect studies need to carefully consider diverse experimental designs and which measurements they can use, their data analyses, and the potential conclusions they might draw.

Further, researchers need to carefully consider the independent and dependent variables, or concepts, that they will be investigating. Vernoy and Vernoy (1992) explain that the independent variable is the factor that is selected, manipulated, or controlled by the researcher; the dependent variable is the measured factor exhibited by the subject.

For example, in studying people's ability to navigate hypertext, Bilsing (1994) hypothesized that subjects' spatial skills and the use of hypertext maps would influence subjects' level of disorientation when navigating a hypertext software program. Spatial skills and the presence or absence of a hypertext map were the independent variables and disorientation was the dependent variable.

Or, in the case of the multimedia research questions, the independent variable is the access to multimedia modules while the dependent variable is the quality of students' laboratory reports.

EXPERIMENTAL DESIGNS

Experiments are the hallmarks of scientific investigation. The researcher develops a theoretical perspective, proposes research hypotheses, and designs an experiment. In this design, the scientist tries to control for all threats to validity, or alternative explanations for the potential findings other than the treatment. In the most simple experiment, a scientist asks, "Does X cause Y?" X represents the possible causative agent (independent variable) and Y represents the potential results of the causative agent (dependent variable). The researcher will measure once, apply X, or causative agent, and then remeasure to see what changes have occurred.

In experimental designs researchers have control over all variables; in quasi-experimental designs they do not have full control over all variables (Campbell and Stanley 1963).

Threats to Validity

Factors that may invalidate the findings are called "threats to validity." Validity centers on whether researchers are measuring what they think they are measuring. Threats to validity are those factors other than the causative factor being investigated that may account for the findings in an experiment. In their classic work, Campbell and Stanley identify eight internal threats to validity resulting from the experimental arrangement and four external threats to the generalizability of the findings.

The eight internal threats to validity include

1. *History*. Something that happens between the measurements may explain the findings. Suppose a researcher was investigating the impact of using computers and word processing software on technical communication skills. During the study, a series of guest lecturers stressed the importance of communication skills to students. Such an event, while not part of the study, could influence students so that they worked harder and thus did better during the study.
2. *Maturation*. During the study, the subjects naturally change in relation to the causative factor—for example, grow older or tired. Suppose you were conducting a usability study to ascertain which way of blocking and moving copy was more efficient. If you had test subjects work for four hours without breaks, fatigue could set in and bias the later results.
3. *Testing*. Testing the subjects at the beginning of the study influences their ability to take the test at the end of the study. For example, simply taking the test at time one helps the subjects do

better on the second test, just as many students do better on the second examination from the same instructor because they learn the instructor's testing methods.

4. *Instrumentation.* The test itself may change between the first measurement and the second measurement. To illustrate, rewording a question could produce drastically different responses; the two sets of data collected would not be comparable.
5. *Statistical regression*, or selecting subjects based on their extreme scores. Assume that in a project to use computers to help students improve their writing, the researchers selected the poorest of the writers. Chances are, the subjects would have improved naturally without having access to computers.
6. *Selection bias.* Using differential selection of the test subjects for different groups may influence the findings. Consider a semester-long writing research project investigating the impact of using computers on students' writing, and all students without computer experience were placed in one group and all students with computer experience in another group. Both groups would then use computers. The differences between the groups could account for differences in the subsequent findings.
7. *Experimental mortality.* Subjects may drop out of one group more heavily than out of another group. On the same research project as above, assume that half of the students with no computer experience dropped out and only 5 percent of the students with computer experience dropped out. Such a high level experimental mortality could invalidate the findings.
8. *Selection-maturation interactions*, or the possible interactions of other factors such as selection and maturation during the study. Again on the semester-long writing research project, assume that students with computer experience improved their writing skills more than students without computer experience. Students coming into the study with computer experience may have matured differently from students without computer experience.

Now consider four external threats to validity that might limit the generalizability of the results.

1. *Reactive or interaction effect of testing.* If testing clues the subjects into what the experiment is about, they may become more sensitive to the experimental treatment. For example, in the research on computers, asking lots of survey questions about writing might signal to students that the study was exploring the impact of computers on writing.

2. *Interaction effects of selection bias and the experimental variable.* The interactions of the selection of subjects and the experimental variable may set up a set of circumstances that produces changes. Suppose that in the writing research project one group of students was selected based on high computer skills, and they were taught to write and compose directly on computers. Suppose the research found that when students with high computer skills were taught specific word processing functions to speed their writing, their writing improved. Those findings cannot be generalized to report that teaching all students the specific word processing functions would improve all students' writing.
3. *Reactive effects of the experimental arrangement.* If subjects know they are part of an experiment, they may act differently. Thus, the findings may not be attributable to the treatment variable under investigation but to the overall experimental arrangement. If in the writing research project one group of students learned that they were being treated differently—for example, they were given laptop computers to use and other students were not—that knowledge could influence their subsequent behavior. Students in experiments may do better because they think they are given special treatment.
4. *Multiple-treatment inference.* The influence of prior treatment may influence a subject's performances on subsequent treatments. Suppose a researcher was investigating students' improved writing skills by considering the variables 1) individual student praise from the instructor, 2) word processing, 3) working in groups in a computer laboratory, and 4) individual conferences with their instructor. The impact of individual praise from the instructor cannot be separated from the impact of the subsequent treatments.

When researchers plan a research project concentrating on cause and effect relationships, they must use specific experimental designs—i.e., the selection of subjects, assignments to different groups, and measurements at different times—to control for validity. Researchers first consider the more likely threats to the validity of their potential findings, and then they develop a research design to minimize those threats.

Researchers have written hundreds of books, chapters, and articles exploring research designs to minimize the threats to validity. Selected works include Campbell and Stanley (1963), Cook and Campbell (1976, 1979); Spector (1981), Lipsey (1989), Brown and Melamed (1990), and Cochran (1992).

An Example Research Design

The following discussion illustrates selected concepts using one of the more powerful research designs—the Solomon Four-Group Design.

Suppose you asked, "Does using a computer with a word processor make students better writers?" The question requires creating a controlled experiment in which you hold everything else constant for four groups except the variables being manipulated—in this case, access and use of word processing programs. You would define a population of students as best you could in terms of age, previous education, previous experience with computers, similar writing skills, and so forth. With that population defined, you *randomly* select and assign the students to one of four groups:

	Time One	Treatment	Time Two
Group 1	O_1	X	O_2
Group 2	O_3		O_4
Group 3		X	O_5
Group 4			O_6

Time one would represent measurements/tests given at the beginning of the semester; time two would represent those at the end of the semester. O_1 represents the measure of Group 1 at the beginning of the semester, O_2 represents measurement of Group 1 at the end of the semester, and so forth. Assume further that the measurements consisted of a survey and a numerical assessment of each student's communication skills. X represents the treatment; in this case, Groups 1 and 3 use a word processor for writing.

Campbell and Stanley (1963) point out that comparing the differences among the scores O_1 , O_2 , O_3, O_4 , O_5, and O_6 provides for replication of the impact of treatment X and controls for the threats to validity of testing at time one as well as the interaction of testing and the treatment. Comparisons of scores O_3 and O_4 might suggest potential changes in subjects that could be attributed to testing or to possible maturation of subjects between time one and time two. A comparison of O_4 and O_5 scores might lend insights into maturation of the subjects.

To develop additional research expertise, turn to the publications cited in the foregoing discussion, or conduct a literature review on experimental design. Books and publications can be found under such general topics as mass communication research, communication research, sociological research, psychological research, educational research, and statistics. Many college and university social science departments offer relevant courses and the faculty in psychology, sociology, journalism, and mass communication departments can help you develop sound research designs.

ADDITIONAL MEASUREMENT TECHNIQUES

The following section briefly highlights additional strategies that can be used to collect data. Keep in mind that triangulation—collecting data by a variety of methodologies—helps minimize the possibility of false conclusions.

Using Scales and Indexes

Scales and indexes enable researchers to ask subjects a series of questions to ascertain what they think about topics, how they think, their learning strategies, and other behaviors. In psychology, social psychology, education, and other related fields, researchers have developed and continue to develop a wide range of scales designed to measure diverse characteristics such as visual processing, field dependence/independence, learning styles, computer phobia, and writing giftedness. Miller (1991) provides detailed discussion and guidance on using more than 50 scales and indexes.

For example, Zimmerman, Muraski, and Peterson (1993) used a professionalism scale and a boundary spanning index, both of which had been developed and used in other studies. Lockwood (1991) used the Mumford-Hayes learning style assessment of students. Moore (1990) used the Kolb learning style assessment. Both scales had been developed by other researchers.

Standard research methods texts for research, mass communication, psychology, sociology, political science, and other social sciences discuss the use of scales and their development. DeVellis (1991) provides a detailed approach that begins with a theoretical discussion of scale development.

Unobtrusive Observations

A fundamental problem plaguing social science research centers on the potential reactive effects of being measured. Simply, when researchers observe, interview, or survey people, the possibility emerges that the measurement may be invalid. No research is without some bias. Bias may come on the part of the researcher or the subjects being studied (Webb et al. 1973).

Researchers may introduce bias through poorly conceptualized, designed, and executed research projects. The subjects being studied may introduce bias through a variety of means: being measured changes people's behavior, some people forget points, some may lie, some may distort the truth, and so on. Thus, careful researchers use triangulation. Still, most measurement techniques involve interacting with the research subjects, and such interactions may bias the results. What can a researcher do?

One way to minimize bias is to not interact with subjects. In their classic *Unobtrusive measures: Nonreactive research in the social sciences*, Webb and colleagues provide a lengthy discussion about using existing information such as physical traces and archival records as well as using simple observation and contrived observations.

Some unobtrusive measures can be quite simple. For example, Colorado State University's M.S. in technical communication is regularly promoted through exhibits at professional communication meetings. To assess interest in the degree program, exhibitors count the number of pieces of literature set out with the exhibit and the number of pieces left after the meeting and subtract the difference. Over three years, between 20 and 30 percent of the people attending the meetings have picked up graduate program information. Such a measure gives a rough indication of interest in that program. Usability studies can also use unobtrusive measurements such as software programs that capture the keystrokes as a person uses an application software. Such programs provide a historical record of the subject's keystrokes, which provides additional data for analyses.

When considering a research project, look closely for potential ways of collecting unobtrusive measurements of the concepts being studied. Such unobtrusive observations can provide additional insights into the research question.

A BRIEF LOOK AT STATISTICS

Quantitative research methods rely heavily on statistics. Statistics can be divided into descriptive statistics and inferential statistics, or sampling statistics (Hsia 1988). Descriptive statistics provide descriptions of the population based on a numerical value. For example, 50 percent of the students favored using personal computers for writing; at the beginning of the semester only 10 percent of the class had experience with desktop publishing software; and so forth.

Inferential statistics, in contrast, require researchers to collect data on a sample of the population, run one or more statistical tests, and then generalize back to the original population. Hsia points out that inferential statistics provide a tool for explaining and predicting selected characteristics of the individuals studied.

Research Questions and Statistical Analyses

For many research projects, researchers investigate the differences among groups, such as treatment and control groups. Other studies track the same group over time by measuring or comparing the group at two or more times on a specific question.

For example, the research project exploring the communication skills of electrical engineers asked the initial research question, "Is there a different in the awareness of the importance of communication among freshman, sophomore, junior, and senior electrical engineering majors?" This question compares four groups. In contrast, a research question that asks, "Does the importance of communication skills to electrical engineering majors increase as they progress in their degrees?" compares the same group. Researchers must measure the same students as freshman, sophomores, juniors, and seniors.

Levels of Data. Statisticians and researchers categorize data as nominal, ordinal, interval, and ratio.

Nominal data constitutes categories by names such as male or female, black or white, freshman, sophomore, junior, senior. Ordinal data constitutes ranking—i.e., placing units in some order such as first, second, and third. Interval data represents data as numerical values with equal intervals between the items, such as 1=strongly agree, 2=agree, 3=neutral, 4=disagree, and 5=strongly disagree. Ratio data also places an equal interval numerical value on data, but includes a base of zero, such as age.

Vernoy and Vernoy (1992) point out that most psychological measurements use either interval or ratio scales, as do reaction times and correct items on a test. They further stress that researchers need to clearly identify nominal and ordinal scales because the statistical analyses required for those are different from the statistical analyses required for interval and ratio data.

Selecting the Appropriate Statistical Tests. For the uninitiated, selecting the appropriate statistics can be confusing. Thus, introductory statistics classes explain the application of statistics to a wide range of topics. You can also gain an understanding of statistics by carefully studying such recent publications as Jaeger (1990), Kanji (1993), Kraemer and Thieman (1987), and Lipsey (1989). Although dated, Frederick Williams (1968) provides a straightforward explanation of statistics. Many social science research methodology texts, such as Babbie (1992), Hsia (1988), Kerlinger (1986), Nachmias and Nachmias (1987), and Singletary (1994), contain detailed discussions that will help the novice develop a clearer understanding of statistics.

While most texts explain statistics and their use, few make clear the connection between the research question and selecting the appropriate statistical tests. For more than 25 years, Richard Powers, University of Wisconsin-Madison, taught a basic statistics course that concentrated on analyzing the research question and selecting the appropriate statistical test.

Powers' approach determines whether the variables produce nominal, ordinal, or interval/ratio data and whether the researcher is comparing independent groups or match groups. Based on this approach, Powers provides a table (table 16.1) to guide researchers to the appropriate statistical tests (Powers 1994).

While Powers provides basic guidance, researchers can select from a host of statistical tests for data analysis beyond those that he recommends. Kanji

(1993) provides succinct summaries of more 100 statistical tests and provides brief explanations of their respective applications. Andrews and colleagues (1978) provide a similar approach that uses a decision tree structure to ask questions about the research question.

Running Statistical Tests

While simple statistics can be run using standard scientific and business pocket calculators, advanced pocket calculators can provide even further analysis. Personal computers can use a wide range of statistical packages, such as SAS, SPSS, and SyStat. Seek out the statisticians, sociologists, psychologists, and communication researchers who can help you select a software package and develop the skills for advanced data analysis. While the learning curve for some software is short, you may encounter problems. Turning to researchers who work with the programs daily can help speed the development of your skills in running the statistical software.

See chapter 13 for an introduction to running statistical analyses on surveys and relevant literature.

TABLE 16.1

A SELECTION GUIDE FOR BASIC STATISTICAL TESTS (POWERS 1994)

	Do TWO groups differ enough to be significantly different?		**Do MORE THAN TWO GROUPS differ enough to be significantly different?**	
	Independent Groups	**Repeated Measurements**	**Independent Groups**	**Repeated Measurements**
Nominal Data	Chi Square Tests	McNemar Test Adjusted McNemar Test Binomial Test Sign Test	Chi Square	Cochran's Q Test
Ordinal Data	Mann-Whitney U Test	Wilcoxon's Test	Kruskal-Wallis Test	Friedman Test
Interval/ Ratio Data	t Test	Paried t Test	Analysis of Variance	Analysis of Variance with Repeated Measurements

REFERENCES

Andrews, F. M., L. Klem, T. N. Davidson, P. M. O'Malley, and W. L. Rodgers. 1978. *A guide for selecting statistical techniques for analyzing social science data.* Ann Arbor, MI: Institute for Social Research, The University of Michigan.

Babbie, E. R. 1992. *The practice of social research.* 6th ed. Belmont, CA: Wadsworth.

Bilsing, L. 1994. The role of external maps and spatial skills in reducing user disorientation in hypertext. Manuscript. Fort Collins, CO, Technical Journalism, Colorado State Univ.

Brown, S., and L. E. Melamed. 1990. *Experimental design and analysis.* Newbury Park, CA: Sage.

Campbell, D. T. 1975. Reforms as experiments. In *Handbook of evaluation research*, ed. E. L. Struening and M. Guttentag. Newbury Park, CA: Sage.

Campbell, D. T., and J. C. Stanley. 1963. *Experimental and quasi-experimental designs for research.* Chicago: Rand McNally.

Christensen, L. B., and C. M. Stoup. 1991. *Introduction to statistics for the social and behavioral sciences.* 2d ed. Pacific Grove, CA: Brooks/Cole.

Cochran, W. G. 1992. *Experimental designs.* New York: Wiley.

Cohen, J. 1977. *Statistical power analysis for the behavioral sciences.* Rev. ed. New York: Academic Press.

Cook, T. D., and D. T. Campbell. 1979. *Quasi-experimentation: Design and analysis issues for field settings.* Boston: Houghton Mifflin.

Cook, T. D., and D. T. Campbell. 1976. The design and conduct of quasi-experiments and true experiments in field settings. In *Handbook of industrial and organizational psychology*, ed. M. Darnette. Chicago: Rand McNally.

DeVellis, R. F. 1991. *Scale development.* Newbury Park, CA: Sage.

Hsia, H. J. 1988. *Mass communication research: A step-by-step approach.* Hillsdale, NJ: Erlbaum.

Jaeger, R. M. 1990. *Statistics: A spectator sport.* Newbury Park, CA: Sage.

Kanji, G. K. 1993. *100 statistical tests.* Newbury Park, CA: Sage.

Kerlinger, F. N. 1986. *Foundations of behavioral research.* New York: Holt, Rinehart & Winston.

Kraemer, H. C., and S. Thieman. 1987. *How many subjects?* Newbury Park, CA: Sage.

Lauer, J. M., and J. W. Asher. 1988. *Composition research.* New York: Oxford Press.

Lipsey, M. W. 1989. *Design sensitivity.* Newbury Park, CA: Sage.

Lockwood, A. 1991. The impact of learning styles on using a desktop publishing system. Manuscript. Fort Collins, CO, Technical Journalism, Colorado State Univ.

Miller, D. C. 1991. *Handbook of research design and social measurement.* Newbury Park, CA: Sage.

Moore, M. B. 1990. The relationship of students' learning style types to their approaches to writing. Manuscript. Fort Collins, CO, Technical Journalism, Colorado State Univ.

Nachmias, D., and C. Nachmias. 1987. *Research methods in the social sciences*. New York: St. Martin's Press.

Powers, R. 1994. *Data analysis in the social sciences*. Madison: Department of Agricultural Journalism, University of Wisconsin-Madison.

Singletary, M. 1994. *Mass communication research*. New York: Longman.

Spector, P. E. 1981. *Research designs*. Newbury Park, CA: Sage.

Stempel, G. H., and B. H. Westley. 1989. *Research methods in mass communication*. New York: Prentice-Hall.

Vernoy, M. W., and J. A. Vernoy. 1992. *Behavioral statistics in action*. Belmont, CA: Wadsworth.

Webb, E. J., D. T. Campbell, R. D. Schwartz, and L. Sechrest. 1973. *Unobtrusive measures: Nonreactive research in the social sciences*. Chicago: Rand McNally.

Williams, F. 1968. *Reasoning with statistics: Simplified examples in communications research*. New York: Holt, Rinehart and Winston.

Williamson, J. B., D. A. Karp, J. R. Dalphin, and P. S. Gray. 1982. *The research craft*. Boston: Little, Brown & Co.

Wimmer, R. D., and J. R. Dominick. 1991. *Mass media research*. Belmont, CA: Wadsworth.

Zimmerman, D. E., M. Muraski, and J. Peterson. 1993. Who are we? A look at the technical communicator's role. Technical report. Fort Collins, CO, Center for Research on Writing and Communication Technologies, Colorado State Univ.

Zimmerman, D. E., M. Palmquist, K. Keifer, M. Long, D. Vest, and M. Tipton. 1994. Enhancing Electrical Engineering Students' Communication Skills—the Baseline Findings. Proceedings of the International Professional Communication Conference 94 IEEE Professional Communication Society.

APPENDIXES

APPENDIX

Accessing Libraries and Databases via the Internet

WHAT IS THE INTERNET?

While all of the talk of a new super information highway caught the public's attention in the summer of 1993, an estimated 15 million people were already using an existing electronic information highway, the Internet, and some 150,000 new users were joining their ranks monthly (Kantor 1994).

The Internet is an outgrowth of the U.S. Defense Department's ARPAnet, developed in the early 1970s to support military research and withstand partial power outages (Krol 1992). Originally, the system consisted of a series of computers linked together by cable. A message could be delivered from user to user quickly over one of several different routes to ensure its prompt delivery.

Today, a series of new networks makes up the Internet, including the National Science Foundation's NSFNET, NASA Science Internet, universities, colleges, government agencies, and businesses (Krol 1992, Kantor 1994). Wilkinson (1993) reports that the Internet consists of more than 45,000 public and private networks. Local area networks are linked to larger networks that are linked to still larger networks. For example, a local area network on one campus is linked to a campus server and several servers are linked to one main computer. In turn, nearby university campuses are linked to a regional supercomputer (Krol 1992, Kantor 1994).

From a user's perspective, you need only concern yourself with accessing the Internet. Today's information seekers gain access through a variety of systems and thereby access a wide range of information.

WHAT INFORMATION IS AVAILABLE?

Electronic mail (e-mail) is perhaps the most popular use of the Internet. E-mail enables users to send letters across the state, nation, or around the world through computer networks.

Of more relevance to researchers, perhaps, are the online public access catalogs. By 1993 at least 280 library catalogs from around the world were accessible through the Internet (Rega 1993). Hundreds of databases, electronic books, journals, newsletters, discussion groups, and related information sources are also available through the Internet. Kantor reported some 4,500 newsgroups covering a wide range of information.

ACCESSING THE INTERNET

To access the Internet, you need a personal computer equipped with either a modem or a network card and the respective software. Modems enable you to connect your computer to a telephone line, as discussed in chapter 5. Universities, government agencies, and companies can also create local area networks by installing network cards into personal computers and cabling the computers with optical fiber to a computer server. With a modem, you either access the Internet through a commercial or no-cost access service.

ACCESS SERVICES

Kantor (1994) provides a review of the primary commercial online services available in late 1993 including America Online, CompuServe Information Service, Dow Jones News/Retrieval with MCI Mail, GEnie, and Prodigy Interactive Personal Services. *PC Magazine*'s Editor's Choice was CompuServe Information Service.

For either no-cost or low-cost access, you may be able to obtain an account on a local academic, government, or business computer system. For some systems you may need to learn the basic UNIX commands. Like DOS for IBM and compatible computers, UNIX serves as an operating system for workstations, RISC, and larger computers. Look for handouts and guidance from the computer support staff, or obtain a copy of *Learning the UNIX Operating System* (Todino and Strang 1990), which provides an excellent introduction to the basics of UNIX.

Other access systems are more user friendly and require only simple steps to move you through a sequence of computer menus. For some academic systems, you need only access Gopher, an interface that provides campus information as well as access to the Internet. Gopher consists of an opening menu followed by submenus that progressively take you into the Internet and the available services. By early 1994, a Windows-based version, WinGopher, provided a Windows interface for accessing Gopher menus.

IDENTIFYING LIBRARIES AND DATABASES TO ACCESS

Identifying available information sources can be difficult, but several publications can help you find the key sources. Mecklermedia Publishing of Westport, Connecticut publishes a series of guides including

Rega, R. 1993. *OPAC Directory 1993: An Annual Guide to Online Public Access Catalogs and Databases.*

Lane, E. and Craig Summerhill. 1993. *Internet Primer for Information Professionals.*

Gregory, Newby. 1994. *Directory of Directories on the Internet.*

Meckler, 1994. *On Internet: An International Title and Subject Guide to Electronic Journals, Newsletters and Discussion Lists on the Internet.*

Kochmer (1992) recommends two online guides for using online public access catalogs:

1. "Internet Accessible Library Catalogs and Databases" by Art St. George and Ron Larsen
   This guide is available through the Internet's FTP (file transfer protocol) from nic.cerf.net and under the directory cerfnet/cerfnet_info and the two files are internet-catalogs-08-91.text for a text only file and internet-catalogs-04-91_apl.ps for an Apple PostScript file.
2. "University of North Texas' Accessing Online Bibliographic Databases" by Billy Barron.
   This guide is available through Internet's FTP from vaxb.acs.unt.edu and under the directory library with file names libraries.ps for the postscript version, libraries.txt for the ASCII text only version, and libraries.wp for the WordPerfect version.

Accessing libraries and databases through the Internet requires having the proper computer and software, knowing how to use your local computer system and its interface, knowing the Internet addresses of the card catalogs and databases you want to access, and learning how to access and use the respective library systems. Each step has numerous pitfalls for the computer illiterate, but once you learn how to use the Internet, you will have a world of useful information available from your personal computer. With the rapid advances in the Internet, look for the latest guides in *Books in Print*. For example, Krol (1992) provided a solid introduction to the Internet, but Hahn and Stout (1994) gave the more recent advancements. Invest time in learning the Internet, and you will reap the benefits.

REFERENCES

Gregory, N. 1994. *Directory of directories on the Internet.* Westport, CT: Mecklermedia.

Hahn, H., and R. Stout. 1994. *The Internet complete reference*. Berkeley, CA: Osborne McGraw-Hill.

Kantor, A. 1994. Internet: The undiscovered country. *PC Magazine* 13(5): 116–18 (March 15, 1994).

Kochmer, J. 1993. *NorthWestNet user services internet guide*. Bellevue, WA: NorthWestNet Academic Computing Consortium.

Krol, E. 1992. *The whole Internet user's guide and catalog*. Sebastopol, CA: O'Reilly & Associates.

Lane, E., and C. Summerhill. 1993. *Internet primer for information professionals*. Westport, CT: Mecklermedia.

Laquey, T., and J. Ryer. 1993. *The Internet companion*. Reading, MA: Addison-Wesley.

Levine, J. 1993. *The Internet for dummies*. San Mateo, CA: IDG Books.

Mecklermedia. 1994. *On Internet 1994: An international title and subject guide to electronic journals, newsletters and discussion lists on the Internet*. Westport, CT: Mecklermedia.

Rega, R. 1993. *OPAC directory 1993*. Westport, CT: Mecklermedia.

Todino, G., and J. Strang. 1990. *Learning the UNIX Operating System*. Sebastopol, CA: O'Reilly & Associates

Wilkinson, S. 1993. Internet steps into the light. *PC Week* 10(43): 105 (November 1, 1993).

APPENDIX B

A Closer Look at Four Database Programs

To prepare this appendix, we tried full versions of Papyrus, Reference Manager, and WPCitation, and a demonstration disk of Pro-Cite. The Pro-Cite demonstration disk was a full working copy but allowed only 30 references to be entered into the database.

Papyrus operates in DOS and can be installed to be accessed through Windows systems and over networks. Papyrus provides an extremely friendly tone in both manuals and software, and the provided directions make installation easy. The opening menu proves easy to follow. To enter a citation, you select one type of template—article, book, chapter, map, patent, thesis, quote, or other. Papyrus automatically provides an abbreviated or extended menu for entering the needed information.

Papyrus has some artificial intelligence that makes entering information easier and quicker. For example, if you enter the name of a journal that has not been entered before, Papyrus queries you, asks for the full journal name, the standard abbreviations, and the data required for citations. Papyrus also has functions that speed entry based on previous information added to the data base. For example, if you enter the abbreviation for a journal previously entered, Papyrus automatically inserts the full name of the journal. Likewise, if you have entered an author's name before, you can type F2 for the glossary, type the first letters of the name, and a possible list comes up. If you highlight the name and do a carriage return, Papyrus automatically inserts the author's full name.

Papyrus includes a series of filters for importing bibliographic files from other reference software and databases, as well as importing searches from CD-ROM databases such as SilverPlatter and Compact Cambridge; online databases such as Medline or DIALOG; Current Contents on Disk or Reference Update bibliographies; dBase programs, and others. Unlike some programs that charge extra for software modules, Papyrus includes the filters free. In addition, you can import existing bibliographies from standard word processing files and convert them into your Papyrus database.

Like other software programs, Papyrus provides an automatic feature that will convert files, marked in the database, to a prescribed stylebook style. It comes with predefined styles or you can define your own style for specific journals.

Papyrus has an easy-to-follow manual and extremely friendly software that carefully walks you through the steps required to produce the needed database and reference list. Rabinowitz (1993) calls Papyrus the value leader of the programs he reviewed, a top-of-the-line program at the baragin-basement price of $99 in 1993.

Papyrus for DOS is available from Research Software Design, 2718 SW Kelly Street, Suite 181, Portland, OR 97201; telephone 503-796-1398; fax 503-241-4260.

Reference Manager Professional Edition Windows was the only Windows-based bibliography software available in late 1993. For users familiar with Windows, the manual's chapter "Learning Reference Manager" walks you through the basics and provides information on entering original data, importing references, retrieving references from a Reference Manager database, creating a transfer file and transferring references, printing alphabetical listing of authors, preparing a bibliography from a manuscript, and preparing a bibliography from a retrieval list.

The program was easy to follow, using the familiar Windows menus and screen structure. The Help menu provided clear guidance to solve the problems we encountered. For the most part, the instructions were clear and the examples provided gave us a general familiarity within an hour or so. The remainder of the manual gives details on the File, Edit, References, and Utilities menus. Appendixes provide detailed guidance on backing up the database, compatibility between the different Reference Manager programs, troubleshooting, Reference Manager files, validating your database, RIS Format specifications, and extended characters.

Reference Manager comes in MS-DOS, Windows, Macintosh, and NEC9801 Series 1MB. The extensive capture module will enable you to convert from a wide range of programs easily and quickly into a Reference Manager database. This feature will enable you to quickly develop databases covering your respective field and to keep the database up-to-date using the update reference software programs. For each system, Research Information Systems markets a Professional Edition Capture module for an additional fee with programs for converting more than 165 different online and CD-ROM databases and a module with formats for about 1,000 different journal citation styles.

Reference Manager is available from Research Information Systems, 2355 Camino Vida Roble, Carlsbad, CA 92009-1572; telephone 800-722-1227 or 619-438-5526; fax 619-438-5573.

WPCitation loads directly into WordPerfect for Windows or WordPerfect DOS and functions using a series of macro commands and the WordPerfect search functions. It does not operate in other word processing programs. When you load WPCitation, the program provides a list of journals from which you mark the ones that you will use. Like many programs, WPCitation will then automatically format the references for the journal style you request.

One template allows you to enter information for articles, book chapters, books, and other references. As you move down the lines (called fields), WPCitation provides a field descriptor in the lower left-hand corner of the screen. The fields include abstract, access phrases for searching, author, role, article/essay title, day/month, descriptors—numbers or form—collection title, editor, issue/edition, journal, keywords, length/comments, pages, place, publisher, series title, translator, volume, and year. Only those fields used appear in the final record (reference) entered in the database.

The 101-page manual provides instructions on installing and running the tutorial, a reference manual, and appendixes. The tutorial begins by having you look at a bibliographic record and then walks you through browsing a data file, generating a bibliography, editing a record, and building a WPCitation data file. It also provides general information. The manual also walks you through building, editing, and using your own WPCitation bibliographic files. The manual details the conventions that guide entering data for a wide range of citations, converting existing citations into WPCitation, and other functions.

WPCitation is available from Oberon Resources, 147 E. Oakland Avenue, Columbus, OH 43201; telephone 614-294-7762.

Pro-Cite provides a demo diskette that allows you to use a sample program of the DOS software and a sample database. The demo disk allows you to explore the full functions of the software, but you cannot generate your own databases. The easy-to-read manual walks you through an easy installation process and the program functions.

Pro-Cite allows you to enter data manually to create files or to download files from CD-ROMs and online data services. To enter data manually, you select the kind of record for the reference, such as journal article, book, or book chapter, and then you enter the reference data by line (fields), such as author, title, date, volume, and page numbers. Pro-Cite treats each reference as a record. Once you have entered the data, you can edit as needed to correct mistakes or to add information.

Pro-Cite enables you to download bibliographies from selected CD-ROM databases and computerized library systems. Some services—BRS, DIALOG, MEDLARS—require the Biblio-Links conversion program to import files.

To search a database, Pro-Cite walks you through identifying the database and using Boolean logic connectors (and, or, not) as well as relational operators (<, <=, =, >=, >) to conduct the respective searches.

As you review your search results, you can mark the individual records, or citations, and direct Pro-Cite to format the references to a particular citation style. If Pro-Cite does not have an automatic formatting program for a style you need, you can create such a program.

Pro-Cite also enables you to sort the files by up to six levels in one pass and to create specific subject bibliographies.

Personal Bibliographic Software, manufacturer of Pro-Cite, packages the software as a full program or a read-only program, and also sells a training kit and training guide. In addition, Personal Bibliographic Software markets Biblio-Links, for downloading files from diverse databases, and Pro-Search, for searching commercial electronic databases.

Pro-Cite, Biblio-Links, and Pro-Search are available from Personal Bibliographic Software, Inc., P.O. Box 4250, Ann Arbor, MI 48106; telephone 313-996-1580; fax 313-996-4672.

REFERENCES

Blumenthal, E. Z., and R. Gilad. 1993. Storing a bibliographic database on your PC: A review of reference-manager software. *The New England Journal of Medicine* 329(4): 283–84.

Neal, P. R. 1993. Personal bibliographic software programs: A comparative review. *BioScience* 43(1): 44–50.

Rabinowitz, R. 1993. Point of reference. *PC Magazine* 12 (17): 269–79 (October 12, 1993).

INDEX

by Linda Webster

XB 2566763 7